A series of student texts in

CONTEMPORARY BIOLOGY

General Editors:
Professor E. J. W. Barrington, F.R.S.
Professor Arthur J. Willis

To Karl and Hally Sax, who first introduced
me to the fascinating study of chromosomes,
this book is gratefully dedicated.

Chromosomal Evolution in Higher Plants

G. Ledyard Stebbins

Professor of Genetics, University of California, Davis

Edward Arnold (Publishers) Ltd., London

© G. Ledyard Stebbins 1971

First published 1971
by Edward Arnold (Publishers) Ltd.,
41 Maddox Street,
London, WIR OAN

Boards edition ISBN: 07131 2287 0
Paper edition ISBN: 07131 2288 9

Printed in Great Britain by
J. W. Arrowsmith Ltd., Bristol

Preface

The purpose of this book is to present a review of the most significant information now available bearing upon the problem of how chromosomal variation between individuals, populations and species contributes to the processes of evolution. It is intended for students who have had some background and training in general genetics and cytogenetics. It should not be regarded as a general textbook in this field. Furthermore, some understanding of the basic processes of evolution: mutation, genetic recombination and natural selection, is assumed. The required background for this book can be obtained from such books as *Chromosome Marker* by K. R. Lewis and B. John; *Cell Biology* by E. D. P. De Robertis, W. W. Nowinski and F. A. Saez; *The Origin of Adaptations* by Verne Grant; and *Processes of Organic Evolution* by the present author.

On the other hand, this book is not intended as an exhaustive treatment of the subject for research workers. From the enormous amount of literature that is available, those references have been selected which are most pertinent to the topic at hand. In the case of controversial points and unproved hypotheses, an effort has been made to present both sides of the case in question. In the interest of economy, references to general or review papers or to comprehensive books have been substituted for references to large numbers of original research papers.

The overall philosophy of this book has been to present evolutionary cytogenetics in terms of relationships between structure and function at the level of the chromosomal organelle. Its basic assumptions are the following. (1) Chromosomes are much more than strings of DNA molecules. They are organelles that, in addition to housing the genetic information, provide mechanisms for releasing and regulating the transmission of this information during development according to a carefully adjusted programme or sequence of events. They also provide for perpetuating and transmitting as units combinations of genes that interact with each other in an adaptive fashion. (2) These complex activities of chromosomes have evolved via processes of mutational change and natural selection because of their adaptive value to the organism or the population in which they exist. In many instances, differences between populations with respect to chromosomal characteristics, as well as all kinds of other characteristics, represent different ways of becoming adapted to similar situations. On the other hand, consistent and wide-spread differences, such as those considered in the present volume, are very unlikely to have evolved in spite of or even without the aid of natural selection, based upon their adaptive value to the populations in which they are found. On the basis of this belief, the unifying thread of relationships between structure and function is regarded as justified.

The author wishes to acknowledge the helpful comments made by a number of cytologists during the preparation of this work, in particular Daniel Zohary, Peter Raven, S. W. Jackson, and Giles Waines. Original photos for some of the illustrations were provided by Ichiro Fukuda, W. M. Hiesey, Marion Ownbey, Giles Waines and Alva Day Whittingham. These are gratefully acknowledged. Most of the manuscript was prepared while the author was a Fellow at the Center for Advanced Studies in the Behavioral Sciences, Stanford, California. The assistance of Center personnel, particularly Mrs. Helena Smith, is also acknowledged with thanks.

Davis, California 95616   G. L. S.
1970

Table of Contents

I

The Three Basic Functions of Chromosomes

Before studying chromosomal variability in relation to the diversity of life which has been the outcome of evolution, we must first become very clear in our minds as to what chromosomes do, and how they perform their functions. This subject is the centre of the discoveries which, since about 1950, have revolutionized our concepts about the nature of life and of heredity. As will be pointed out later in this chapter and in various places throughout this book, many of the problems concerning the functions of chromosomes are still unsolved, and are the subject of considerable controversy. Nevertheless, the tools for solving most of these problems are already available, so that their solution depends in large part upon the ingenuity of biologists in designing the right experiments, and their persistence in carrying such experiments through to completion. Consequently even in a book like the present one, which is designed for students who are beginning their careers in the life sciences, and are primarily interested in understanding evolution and its outcome in terms of the diversity of plant life, a discussion of these problems is not out of place.

DNA AND ITS REPLICATION

The first and primary function of chromosomes is *storage, replication and transmission of hereditary information*, contained in the genes. Replication makes possible the existence of a complete set of genes in every cell nucleus of the body, including those in the reproductive organs which give rise to the gametes. The familiar process of cell division by *mitosis*, at which chromosomes become visible in the forms usually illustrated, is in all somatic or body cells preceded by gene replication. The only exception to this rule occurs between the two modified mitotic divisions which constitute the process of *meiosis*, which brings about the reduction in chromosome number prior to the formation of spores, gametophytes, or gametes. Since no gene replication occurs between the two meiotic divisions, the split halves of the chromosomes become segregated at these divisions in such a way that each resulting nucleus contains only half as many chromosomes and genes as did the original nucleus at the beginning of meiosis. The transmission of information is complete when two gametes, each possessing the reduced or haploid number of chromosomes, unite to form the zygote, and so restore the original diploid number of chromosomes and genes in the body cells of the offspring.

The chemical structure and replication of DNA

The essence of replication is the distribution of an identical code of genetic information to every cell of the body. This is made possible by the structural nature of the molecules which contain this information, deoxyribose nucleic acid, or DNA.

The chemical structure of DNA is now explained in so many books, including elementary textbooks of biology, that an additional explanation here does not appear to be necessary. The following features of its structure are, however, essential to an understanding of the role of chromosomes in evolution. In the first place, the precise hydrogen bonding between the base pairs adenine–thymine and guanine–cytosine makes possible the exact duplication of long and complex sequences of nucleotides. Secondly, gene specificity depends upon the presence of specific ordered patterns of nucleotides, only four kinds of bases being present. The 'telegraphic message' contained in the order of base pairs of a single gene always includes hundreds of units, and may include thousands of them. If at any one of these hundreds or thousands of sites within the gene one kind of base pair is substituted for another, a mutation has occurred. This means that any gene can mutate in thousands of different ways. Each variant of the order sequence of nucleotides present at a single gene locus constitutes a different allele of that locus. Hence in a large population of individuals, the number of possible order sequences, and hence of different alleles, is astronomical. Since four possibilities exist at each nucleotide site, and each of these can be combined in four different ways with nucleotides at any other site, the number of possible alleles of any gene is 4^n, where n is the number of nucleotides of which the gene consists.

Multiple alleles are now well known to geneticists, but only with the advent of recent knowledge concerning the nature of the gene has the potential variability of a single gene been fully appreciated. Even if there is less than one per cent of the variant sequences code for functionally adaptive proteins, the number of possible functional alleles at any one locus is nevertheless large.

The replication of the DNA molecule can begin only when the cell nucleus in which it exists contains free nucleotides in a particular form. This is known as the phosphorylated state. It involves the attachment to the nucleotide of two additional phosphate groups by means of special chemical bonds, which are often called 'high energy' phosphate bonds. The wide-spread function of these bonds in accepting and releasing energy is well known to modern students of elementary biology. The fact that nucleotides are phosphorylated before being incorporated into DNA means that the incorporation process itself does not require an extra expenditure of energy.

On the other hand, since replication cannot begin until phosphorylated nucleotides are available, the chemical synthesis of the nucleotides, as well as their phosphorylation, are essential preludes to the process of replication itself. Synthesis and phosphorylation of nucleotides can be carried out only with the aid of a battery of enzymes, each of which catalyses specifically only one step in the biosynthetic pathway required for their synthesis.[83] Each of these enzymes is coded by a different gene, which must be activated to produce the necessary enzyme before replication can begin. At least one of these enzymes, thymidine kinase, which catalyses the phosphorylation of thymidine, has been shown by means of critical experiments to be a primer for replication in plant tissues. Consequently, it is highly probable that the timing of replication, and consequently of nuclear division, is in many instances controlled by regulating the action of the genes which control the enzymes involved in the synthesis and phosphorylation of nucleotides, as well as those concerned with the replication process itself.

These facts are mentioned to emphasize the fact that, contrary to the impression which many biologists have gained, DNA is not a self replicating molecule. Removed from the cell nucleus, and placed in a test tube, even in the presence of free nucleotides of the right kinds, but without the necessary enzymes, and lacking adenosine triphosphate (ATP) as a source of energy, DNA is no more self replicating than is a copper wire. DNA is only an informational template for a self replicating system. Given conditions under which this self replication can take place, the molecular structure of DNA is such as to assure the fact that the replicate will have the same informational content as the original.

The process of replication itself takes place as follows. With the aid of specific enzyme, the double helix becomes unwound over a part of its length, and the hydrogen bonds joining the complementary bases in this region are broken, so that the two strands become separated from each other. This makes possible the entrance of free nucleotides and their attachment by means of newly formed hydrogen bonds to complementary bases of the two existing strands. Another enzyme then catalyses the formation of new covalent bonds between the sugar and phosphate groups of the previously free nucleotides, thus forming for each pre-existing helix of DNA a new, complementary one.

This form of replication is called *semi-conservative*, since it involves the conservation of one of the two strands, and the formation of a new one beside it. That this is the actual method of replication was demonstrated by J. H. Taylor in 1958, by the use of radioactively labelled nucleotides. He reasoned that if such nucleotides were permitted to enter the cell during only one cycle of replication, their position in the replicated strands, as well as in the strands of daughter chromosomes formed after later cycles of replication, would provide evidence about the naure of replication. He found that chromosomes formed after one cycle of replication, during which labelled nucleotides were introduced, contained one unlabelled strand, the one conserved from previous replications, and one labelled strand, formed by the labelled nucleotides which had been introduced. If the cells were permitted to pass through two or more divisions after the labelled nucleotides had been introduced, the number of unlabelled strands, which were formed from free nucleotides synthesized after the introduction of the labelled ones, increased proportionally, according to expectation on the basis of the semi-conservative replication hypothesis.

This account of replication can be summarized by saying that the double helix of DNA is a semi-conservative **template,** which is able to produce an indefinite number of exact copies of its pattern because of the precise hydrogen bonding between complementary bases: adenine with thymine and guanine with cytosine.

The commonest form of mutations, which are the building blocks for the great bulk of evolutionary change, consists of rare mistakes in the replication process. One way in which mutations can occur is through mismating between bases during one cycle of replication. If, through a rare chemical accident, adenine becomes hydrogen bonded with cytosine rather than with thymine during one replication, then the cytosine will at the next replication attract its normal partner, guanine. The latter base will thereafter form a permanent unit of the template, in place of adenine, at the site where the 'mistake' occurred.

REGULATION OF GENE ACTION

The second function of chromosomes in higher plants and animals is to control gene action in such a way that the primary products of the genes are released at the right times and in the right amounts to produce an orderly and specific sequence of biochemical and cellular events during development. Until recently, biologists often asked the question: 'How is it that the different cells of the body, all of which contain the same genes, can become as different from each other as parenchyma cells, root hairs, the vessel and fibre cells of wood, and all of the other kinds of cells which exist side by side in the same plant?' In its broadest outlines, the answer to this question can now be given. In most tissues, particularly those undergoing the last stages of specialized differentiation, only a small proportion of the genes are actively producing their primary products.[164] By means of various kinds of regulating agents, genes can be 'turned on' and 'turned off' in ways which adjust their functions to a sequence of harmonious, delicately balanced interactions with each other and with various factors of the cellular environment. Moreover, the action of at least some genes which are 'turned on' is not constant in intensity at all times, but may be amplified or reduced in response to various kinds of signals. The nature of the regulators which control gene action is still largely unknown, but some of the information which we have about this subject is presented in the next chapter. First, however, we must review the processes involved in primary gene action itself.

The transcription of RNA

The transcription of RNA resembles in many respects the replication of DNA. It also requires a series of reactions which are catalysed by different enzymes. The nucleotides which will form the RNA, like those of DNA, must be phosphorylated before transcription itself can begin. The order of nucleotides in the RNA molecule is determined by the order already present in the DNA molecule, based upon the principle of complementary hydrogen bonding between bases: uracil binds to adenine while cytosine binds to guanine. After they have become bound to the DNA by means of hydrogen bonds, the ribose nucleotides of RNA become covalently bound to each other by means of the sugar and phosphate groups which form the 'backbone' of the molecule. Once this is formed, the RNA becomes freed from its DNA template through release of the hydrogen bonds.

Translation and the synthesis of proteins

The synthesis of proteins requires three different kinds of RNA: *ribosomal, transfer,* and *messenger.* Their molecules differ greatly from each other in respect to size and conformation, and play completely different roles in the translation process. The information that is coded in the gene is transcribed onto the messenger RNA, which consequently has molecules of different sizes, depending upon the length of the 'message' being carried. Transfer RNA provides the link between the nucleotides of RNA and the amino acids which are the units of the protein molecule. Its relatively small molecules are of more than twenty different kinds, one or more of which can become chemically bound to each one of the twenty kinds of amino acids which are incorporated into proteins. The binding of a particular kind of transfer RNA to one kind of amino acid is catalysed by a specific activating enzyme. This enzyme determines that a particular kind of transfer RNA can become bound to one and only one kind of amino acid. We can thus speak of valyl transfer RNA, which is specific for valine, seryl tRNA, specific for serine, etc. Each of these bears at a specific position on its helix three bases which are complementary to a triplet of bases present on the messenger RNA. These are known as the **codon** (on the messenger) and the **anticodon** (on the transfer RNA). The relationship between codons and amino acids is a complex one, embodied in the well-known genetic code. It is, however, very precise, so that the linear arrangement of bases on the messenger RNA corresponds directly to the linear arrangement of amino acid residues in the protein chain.

The process of translation requires still another kind of RNA known as **ribosomal RNA.** It is produced by the DNA template existing in one small part of the chromosomal complement, that which is associated with the nucleolus, a small organelle in the nucleus which will be described and illustrated in the next chapter. Ribosomal RNA becomes complexed with structural proteins to form small organelles known as *ribosomes*, to which the messenger RNA molecules become attached during the process of translation. As a matter of fact, the great bulk of RNA existing in a cell at any one time is ribosomal RNA. This is partly because the DNA regions which code for it are at certain stages of development more active in transcription than are the much more numerous regions which code for messenger RNA. The chief reason for the greater quantity of ribosomal RNA is, however, the relative permanence of its molecules. Most of the molecules of messenger RNA in actively metabolizing cells disintegrate after having performed their function for a few minutes or hours; the RNA–protein complexes which form ribosomes break up much more slowly. Some of them may accept a succession of different kinds of messenger RNA, one after the other.

One might imagine that a complex process like translation could be regulated in a number of different ways. A variety of experimental evidence suggests that this is the case. In some instances, translation appears to be restricted by limiting the number of ribosomes to which messenger RNA can become attached. At other times, the availability of suitable transfer RNA's appears to be the limiting factor. There is, however, evidence to suggest that at least in bacteria the rate of the translation process itself is remarkably constant. The amount of a particular kind of protein which can be synthesized at any one time can be increased only by increasing the number of ribosomes and of messenger and transfer RNA molecules present in the cell, so that several protein chains can be synthesized at the same time.

This description of transcription, translation, and protein activation has been presented in order to emphasize the point that the activity of genes and their immediate products can be regulated in various ways and at various stages in their formation. Since chromosomes contain DNA and therefore must be the sites of transcription, we might expect that chromosomal regulation of gene activity acts chiefly or entirely at the level of the transcription process. This assumption is probably correct. Consequently in the present book, which will discuss only the roles of the chromosomes themselves in heredity and evolution, only the regulation of replication and transcription will be taken up in later chapters. Nevertheless, one should realize that many other ways of regulating gene action exist, and under many conditions these extra-chromosomal mechanisms may be the chief ones which are operating.

CHROMOSOMES AS REGULATORS OF GENE RECOMBINATION

The third function of chromosomes is to regulate gene recombination in the segregating progeny of a hybrid between two genetically different individuals.

An axiom of genetics is that genes which are located on different chromosomes segregate according to the principle of independent assortment (Mendel's second law), while genes located on the same chromosome are genetically linked, so that combinations of alleles which were present in the parental genotypes appear more frequently in progeny of hybrids than do new gene combinations. Linkage is, therefore, a conservative force. Its strength depends upon two factors. One of them is the distance between two gene loci on the chromosome, and the other is the frequency of crossing over, which provides recombination in spite of linkage, in the region between the two loci concerned. Obviously, therefore, the total amount of genetic recombination which is possible in a hybrid between two widely different genotypes depends upon (a) the number of chromosomes and hence of linkage groups possessed by the species, and (b) the presence or absence of various kinds of factors, including some specific genes, which regulate the amount of crossing over.

The importance of linkage and crossing over lies not only in their effects upon the total amount of genetic recombinations possible in a genetically heterogeneous and cross fertilizing population, in which a large amount of segregation is taking place. In addition, linkage may serve as a device for keeping together groups of genes which cooperate with each other to produce some particularly adaptive combination of characteristics. A good example of such an adaptive combination is the series of genes responsible for heterostyly in the genus *Primula*. As was first discovered by Charles Darwin, primrose flowers are of two kinds (Fig. 1.1). They differ from each

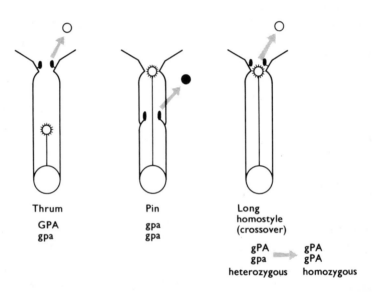

Thrum	Pin	Long homostyle (crossover)
GPA	gpa	
gpa	gpa	

gPA ⟶ gPA
gpa   gPA
heterozygous   homozygous

Fig. 1.1 Diagrammatic vertical sections of the two kinds of flowers, thrum and pin, which are found in a normal heterostylous species of *Primula*. Below each flower is the genic constitution as postulated by Dowrick on the basis of the linked gene cluster hypothesis. At right, self fertilizing long homostyle flowers, which are produced by crossing over between G and P.

other in respect to (1) the length of the style, (2) the position of the anthers on the corolla, (3) the nature of the papillae on the surface of the stigma, (4) the surface of the pollen grains, and (5) certain immunological characteristics of their proteins which favour pollen tube growth in 'legitimate' crosses, i.e. between long-styled and short-styled genotypes, and hinder it in crosses between similar genotypes.

Genetic studies[57] have shown that, in general, these five characteristics are inherited as a single Mendelian unit. When long-styled plants, known as pin, are crossed with short-styled plants, known as thrum, the offspring always segregate in a 1 : 1 ratio of pin to thrum. Under exceptional conditions, either kind of plant can be self fertilized. In that case, pin plants always breed true for all of the pin characteristics, while selfed thrum plants produce progeny in the ratio 3 thrum to 1 pin. We can therefore designate the pin 'allele' by the symbol s and regard thrum as determined by its dominant 'allele' S, the respective genotypes being ss for pin and Ss for thrum.

Large scale genetic experiments have, however, shown that S and s are not alleles of a single gene. Occasional plants arise from the progeny of crosses which have some characteristics of thrum and others of pin.[57] Analysis of these plants suggests strongly that they are the result of rare crossing over between tightly linked genes. The symbols S and s, therefore, do not stand for alleles at individual loci, but for groups of closely linked alleles. Geneticists have given the name **supergene** to such clusters of closely linked genes which are inherited as a single unit.

When we consider the evolutionary history of heterostyly in *Primula*, we are forced to recognize that the gene cluster responsible for this syndrome of characteristics must be extremely old. The morphological characteristics which it produces are found in scores of species of *Primula* distributed throughout the northern hemisphere. The probability that such an unusual group of characteristics arose and became associated with each other in the same way more than once is so low that it can be disregarded. Hence we are forced to conclude that this particular cluster of contrasting alleles has persisted unchanged throughout the tens of millions of years which must have been required for the evolution of the diverse species of *Primula*. Mutations which would have altered the genes of the cluster, or chromosomal changes which would have separated its genes, must certainly have occurred, but must usually have been rejected by natural selection, because of their adaptive disadvantage. There must have been some exceptions to this rule, since a few species of *Primula*, which for various reasons are believed to be derived, recently evolved types, have flowers of only one kind, and are largely self pollinating.

Very pertinent to the subject of the relation between the role of chromosomes as regulators of recombination and the evolutionary significance of variations in gross chromosome structure is the question: 'How did the gene cluster responsible for heterostyly in *Primula* originate?' The possibility that exactly the right mutations would occur in genes which happened to exist next to each other on the chromosome is very remote, when we recognize the fact that the chromosomal complement as a whole must contain thousands of genes with very diverse functions. On the other hand, we can imagine without too much difficulty a situation in which two of the characteristics involved in heterostyly were coded by genes lying on the same chromosome, and hence linked, but with an appreciable amount of crossing over. Such a system would work, though rather inefficiently, through successful cross pollinations chiefly between non crossover individuals, the products of crossing over being less efficient reproducers. Given such a situation, a chromosomal rearrangement, such as an inversion of a segment of the chromosome, which would place the cooperating genes closer to each other and would reduce the amount of crossing over between them, would have a decided adaptive advantage and would quickly spread through the population in which it might occur. Moreover, if by mutation a gene located on a different chromosome should produce an allele capable of reinforcing the heterostyly, it would have a good chance of being preserved and spreading through the population. Once it had become established in the population, translocations between non-homologous chromosomes which would place it near the other genes responsible for heterostyly would have such a high adaptive advantage that one would expect them eventually to occur and become established.

This reasoning is the basis of one of the most important hypotheses for explaining the fact that, as outlined in Chapter 4, structural rearrangements of chromosomes have played an important role in the evolution of plant species. According to this hypothesis, which can be called the *adaptive gene cluster hypothesis*, constant differences between populations and species in respect to such chromosomal rearrangements as inversions and translocations have become established chiefly when they have conferred an adaptive advantage by bringing together into linked clusters groups of genes which cooperate to produce some highly adaptive combination of characteristics. This hypothesis will be discussed more fully, with examples, in Chapter 4.

Alteration of recombinations by tetrasomy

Another way in which chromosomal changes can affect genetic recombination is by duplication of gene loci, so that four alleles of a particular gene are found in a single plant rather than the usual number of two. This is known as the *tetrasomic* condition. It happens most often when the entire chromosomal complement of a species is doubled to produce the *tetraploid* condition. It can also be produced by duplication of individual chromosome pairs, or even of parts of chromosomes.

The effect of the tetrasomic condition on the genic constitution of a hybrid can be understood most easily by following the segregation and recombination of alleles in a cross between two parents which differ from each other at a particular gene locus with respect to four alleles, rather than two. We make the two assumptions: (1) each allele is located on a different chromosome, and (2) the chromosomal regions segregate at random, at the first division of meiosis. We then get the following situation:

Parents: AAAA ×   aaaa

Gametes: AA   aa

F_1 somatic cells: AAaa

F_1 gametes: 1/6 AA, 4/6 Aa, 1/6 aa

F_2 genotypes: (obtained by squaring the expression in the line above)
                      1/36 AAAA, 8/36 AAAa, 18/36 AAaa, 8/36 Aaaa,
                      1/36 aaaa

This calculation tells us that the F_2 progeny of a mono-hybrid cross, which in the usual disomic or typical Mendelian ratio of 1 : 2 : 1 contain 50 per cent of parental genotypes and 50 per cent of intermediates, in the tetrasomic ratio contain only 5.5 per cent (1/18) of parental genotypes and 94.5 per cent of intermediates. In the case of alleles of which neither one is dominant over the other, this difference affects greatly the appearance of natural populations. Tetrasomic inheritance has a buffering effect on intermediate genotypes which can be of great adaptive value to a population if such genotypes confer hybrid vigour, or a superior adaption to a particular environment. This subject will be taken up again in Chapter 5, in connection with the phenomenon of polyploidy.

TYPES OF CHROMOSOMAL VARIATIONS

Cytologists who compare the chromosomes of related species can recognize five different kinds of variations, as follows.

1. *Variations in absolute chromosome size*

Chromosome size, including the total DNA content of the nucleus, may vary as much as 20-fold between genera of the same family having the same or similar basic chromosome numbers (Fig. 1.2). When vascular plants are compared with mosses, algae, fungi, and various kinds of microorganisms, vascular plant chromosomes are found, in general, to be larger than those of lower forms. This indicates that, at lower levels of structural complexity of the plant, the prevailing evolutionary trend has been toward increase in both chromosome size and nuclear DNA content. On the other hand, this trend cannot be observed when different groups of vascular plants are compared, since advanced genera belonging to highly specialized families such as the Cruciferae (*Arabidopsis*), Gramineae (*Panicum*) and Gesneriaceae (all genera) have smaller chromosomes and, where measured, lower nuclear DNA contents than primitive vascular

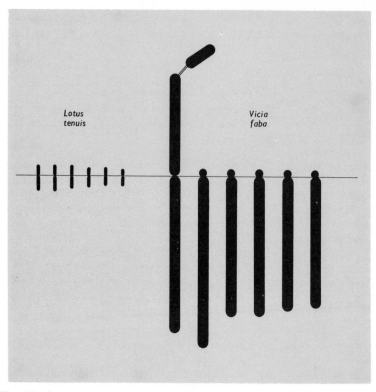

Fig. 1.2 Contrasting chromosome sizes in two species belonging to the same family (Leguminosae) and having the same basic number, $x = 6$: *Lotus tenuis* and *Vicia faba*.

plants such as *Psilotum, Tmesipteris, Lycopodium,* and *Botrychium.* In different groups of flowering plants, trends toward increasing chromosome size as well as toward decreasing size can be detected. These trends and their possible significance will be discussed in Chapter 3.

2. *Variations in staining properties*

Cytologists are familar with the fact that after similar fixation techniques, and exposure to the same stains at the same concentrations, the chromosomes of different species vary greatly in their ability to adsorb the dyes on their surfaces. Chromosomes of most Liliaceae and Gramineae are notable for the ease with which they can be stained; those of Malvaceae and Onagraceae stain with much greater difficulty. In addition, the chromosomes

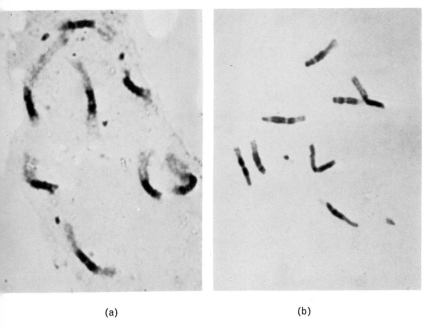

(a)   (b)

Fig. 1.3 Somatic chromosomes at (a) early and (b) late prophase in root tips of *Plantago insularis* ($2n = 8$), a species having pronounced heterochromatic regions near the centromere.

of many species contain some regions which stain more easily than others at many stages of the mitotic cycle. These have been designated *heterochromatic regions.* They are illustrated in Figure 1.3. Although the reasons for these differences in staining properties are by no means clear, they are probably not associated with differences in the molecular structure of the DNA. On the other hand, the reasons for these differences are probably to be sought in the molecular composition and the relationship to the DNA of the various kinds of proteins which exist in all chromosomes except for those of bacteria, blue-green algae, and viruses. These differences in staining properties and their possible significance are discussed in the next two chapters.

3. *Variations in chromosome morphology*

Chromosome morphology is usually studied at the metaphase of mitosis, when chromosomes have become contracted to the maximum amount or nearly the maximum in their cycle, and when they are most easily stained. The principal landmarks which may be seen at this stage are the **centromere** (sometimes known as the kinetochore), to which the spindle fibres are attached, and in many chromosomes one or more **secondary constrictions.** In addition, one, two, and (rarely) a larger number of chromosome pairs in the somatic complement of a species, bear at one end a **satellite.** This usually appears as a single small spherical body, or a pair of such bodies, attached to the remainder of the chromosome by a slender thread. These chromosomal landmarks are illustrated in Figure 1.4.

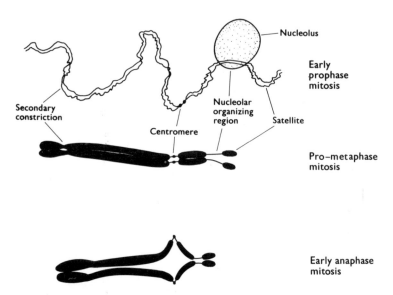

Fig. 1.4 Diagrams showing the landmarks or distinguishing features of a somatic chromosome at early prophase, pro-metaphase and early anaphase of mitosis.

In most chromosomes, the centromere is localized in one particular region of the chromosome. According to its position, chromosomes are designated as follows (Fig. 1.5):

telocentric: centromere at one end of the chromosome, so that the chromosome consists of a single arm.

acrocentric: centromere near one end of the chromosome, so that it contains one long arm, and one very short arm, which is nearly or quite isodiametric.

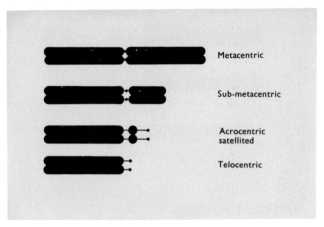

Fig. 1.5 Diagrams showing the appearance of somatic chromosomes having centromeres at different positions.

sub-metacentric: centromere nearer to one end of the chromosome than the other, so that the two arms are distinctly unequal, but less so than in acrocentric chromosomes.

metacentric: centromere at or near the middle of the chromosome, so that its arms are nearly or quite equal in length.

These four categories are not sharply distinct, but grade imperceptibly into each other. They provide useful terms in the case of many complements having chromosomes with radically different morphologies. In other instances, however, the chromosomes belonging to a particular complement are best characterized by obtaining quantitative figures for the ratio between the lengths of their two arms.

Some species of plants, particularly in the families Juncaceae (rushes) and Cyperaceae (sedges), have chromosomes with diffuse centrometric regions which cannot be classified according to the categories mentioned above. Chromosomal variation in these plants is discussed in Chapter 4.

Telocentric chromosomes can arise from meta- or acrocentric chromosomes by means of breaks in the centromeric regions, followed by a sufficient amount of synthesis of new centromeric material so that the half centromeres which are produced by the breakage can resume their original function. Similarly, metacentric chromosomes can arise by the fusion of two telocentrics, without any material alteration of either the chromosomal contents or the arrangement of the genes. These changes are apparently very common in the evolution of animal species. In plants, however, telocentric chromosomes are rare. Their occurrence in angiosperms and

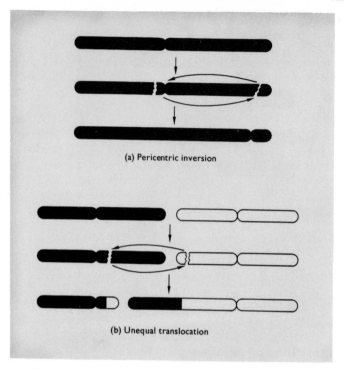

(a) Pericentric inversion

(b) Unequal translocation

Fig. 1.6 Diagrams showing how the position of the centromere can be altered by **(a)** a pericentric inversion and **(b)** an unequal translocation.

their relationships to metacentric chromosomes will be discussed further in Chapter 4.

Alterations in the position of the centromere which convert metacentric to acrocentric chromosomes and *vice versa* are accomplished either by inversions of chromosomal segments which include the centromere, i.e. ***pericentric inversions***, or by interchanges of unequal chromosomal segments between non-homologous chromosomes. Figure 1.6 shows how this can arise. This phenomenon is discussed further in Chapter 4.

4. *Variations in relative chromosome size*

The chromosomal complements or **karyotypes** of most species of plants consist of chromosomes which are comparable to each other in size. There are, however, many complements which contain chromosomes of two contrasting sizes, large and small. A striking example is that of the related genera *Yucca* and *Agave* in the Agavaceae (Fig. 4.13). In other instances, as in *Delphinium* (Fig. 1.7b), a graded series of chromosomal sizes exists, as is true of many animals, including man. Compared with related species having homogeneous karyotypes, those having strongly heterogeneous karyotypes usually exhibit a number of specialized morphological characteristics. In general, homogeneous karyotypes are regarded as more primitive, and heterogeneous karyotypes as more specialized. This topic is discussed further in Chapter 4.

Many species of plants contain, in addition to the normal, constant complement of chromosomes, a variable number of chromosomes which are much smaller and are usually heterochromatic. These chromosomes, known as accessory or B chromosomes, are distinguished from the smaller chromosomes found in such genera as *Yucca* and *Agave* by their tendency

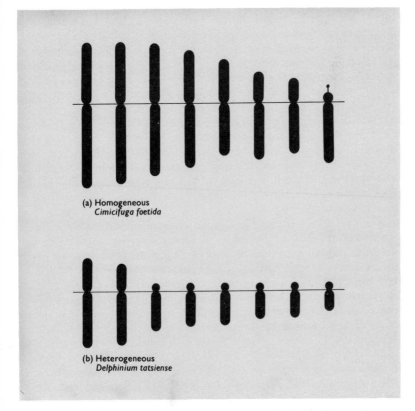

(a) Homogeneous
Cimicifuga foetida

(b) Heterogeneous
Delphinium tatsiense

Fig. 1.7 Karyotypes of two species belonging to the family Ranunculaceae. **(a)** The homogeneous karyotype of *Cimicifuga*. **(b)** The heterogeneous karyotype of *Delphinium*. (From Levitzky.[134])

to vary in number, usually between different individuals of the same population, and often between different tissues of the same individual. Moreover they differ from the larger chromosomes of the same complement in their staining properties, while the smaller chromosomes of normal complements resemble the larger ones in this respect. The evolutionary significance of accessory chromosomes is discussed in Chapter 3.

5. *Aneuploid variations in chromosome number*

Variations in chromosome number are of two very different kinds, **aneuploid** and **polyploid** or **heteroploid.** Aneuploid variations form series in which the gametic numbers of related species form consecutive series, or more rarely they differ from each other by two or more. Thus the gametic numbers of most species of the genus *Crepis*[6] form the series $x = 3, 4, 5, 6$, and 7, while in *Crocus*[42] every number from $x = 3$ to $x = 15$ is represented by at least one species. Evidence of various kinds, such as degree of specialization in external morphology, geographical distribution, and chromosome numbers in related groups, indicates that trends of evolution have gone in different directions in different groups. Thus in *Crepis*[6] the most primitive basic number was probably $x = 6$, so that the series is a descending one, except for the rise from $x = 6$ to $x = 7$ in one small section of alpine and subalpine species. In *Clarkia*,[139] on the other hand, the series is chiefly an ascent from $x = 7$ to $x = 8$ and $x = 9$, though in one section of the genus the descent to $x = 5$ has taken place. In *Dorstenia*[209] (Moraceae), the most primitive number is probably $x = 14$, which in the Old World species gave rise to a descending series through $x = 12$ to $x = 10$, while the New World species form an ascending series to $x = 15$ and $x = 16$.*

As is explained more fully in Chapter 4, aneuploid series are usually the by-products of unequal translocations between non-homologous chromosomes. In descending series, the loss of genetic material which accompanies the reduction in number is confined to a single centromere and the immediately adjacent chromosomal regions, which apparently become genetically inert before the loss takes place.[20] Aneuploid increase can be achieved through the formation of centric fragments, consisting of extra centromeres plus small chromosomal regions next to them. Such fragments appear not infrequently in the progeny of structural heterozygotes for radiation-induced translocations, and their occurrence in structural heterozygotes of natural origin is to be expected. Centric fragments are usually lost in later progeny of the plants concerned, but, if they should receive by translocation a portion of the arm of another chromosome containing

* The symbols adopted by cytologists in referring to chromosome numbers are as follows: N = gametic number, $2n$ = somatic numbers of a species, x = basic gametic number or numbers of a genus.

active and essential genetic material, they could form the basis for phylogenetic increase in basic number.

6. *Euploid and heteroploid variations in chromosome number*

By far the commonest variations in chromosome number in vascular plants are doublings and higher multiplications of entire chromosome sets. In many plant genera all of the species have numbers which are multiples of the basic gametic (often called the 'haploid' or 'monoploid') number for the genus. Thus in *Agropyron* there exist species with somatic numbers which are multiples of 7, i.e. 14, 28, 42, 56, and 70, while in species of *Chrysanthemum* there exists a series based upon 9, i.e. 18, 36, 54, 72, and 90. The overwhelming body of evidence in the case of most such *polyploid series* is that the lower numbers are primitive and the higher ones are derived, but occasional reductions from higher to lower levels are not impossible. The significance of such series is discussed in Chapters 5 and 6.

Polyploid series which consist entirely of multiples of a single basic number are designated as *euploid.* In many genera, however, similar series have included species derived by chromosome doubling from hybrids between diploid species having different basic numbers. Thus among the cultivated species of *Brassica*, *B. napus* has the somatic number 38, having been derived by chromosome doubling from a hybrid between *B. campestris* ($2n = 20$) and *B. oleracea* ($2n = 18$).[209] Since another species, *Brassica nigra*, has $2n = 16$, a combination of chromosome doubling, involving individual species, hybrids between species having the same chromosome number, and hybrids between species having different numbers, has produced in this genus, in addition to the original aneuploid series including the gametic numbers $n = 7$, 8, 9, and 10, a heteroploid series including $n = 16$ (somatic number $2n = 32$), 17, 18, 19, 20 ($2n = 40$). A much more extensive series existing in the genus *Stipa* and its relatives (*Oryzopsis*, *Piptochaetium*), [42] and including the gametic numbers $n = 11$, 12, 14, 16, 17, 18, 19, 20, 22, 23, 24, 30, 32, 35, and 36, has probably a similar origin. Figure 1.8 is a generalized diagram showing how such numbers could arise.

Heteroploid series having a different origin exist in a few plant genera. They are of two kinds. One is derived from the occasional ability of hybrids having uneven somatic numbers to produce viable progeny constituting a series of different numbers. For instance, hybridization between a species having $2n = 28$ and another having $2n = 14$ will produce a triploid F_1 with $2n = 21$. Its progeny may include plants with $2n = 21$, 22, 23, 24, 25, 26, 27, and 28. If the species are sexually reproducing, plants having such deviant numbers are usually sterile and quickly disappear from the population. If, however, asexual reproduction has

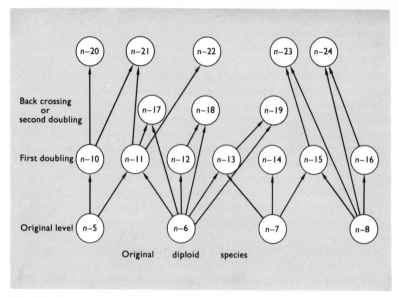

Fig. 1.8 Diagram to show how successive hybridizations and chromosome doublings can lead to an extensive series of aneuploid chromosome numbers. Each of the higher numbers can be derived from various combinations of lower numbers, only one of which is shown.

become predominant in the genus, these numbers may be perpetuated indefinitely. A good example of this situation is found in the genus *Poa*.[209]

Another kind of heteroploid series exists in genera having chromosomes with diffuse centromeres. Such chromosomes can break at many points along their length and give rise to two smaller chromosomes which can still become oriented on the mitotic spindle and carry out the normal succession of splitting and separation of daughter chromosomes. If fragmentation of this type has become combined with translocations of chromosomal segments, by hybridization between species having different numbers, and chromosome doubling in the progeny of such hybrids, extensive aneuploid or heteroploid series can be built up. This is the probable origin of the series found in various genera of Cyperaceae, particularly *Carex*.

CHROMOSOMAL DIFFERENCES AS TAXONOMIC CHARACTERS

Some taxonomists have asserted that chromosomal differences represent just another morphological character and should be treated in the same way as the various characters of external morphology. If characters are to be regarded only as means to the end of fitting individuals into a logical hierarchy of species and higher categories, and of identifying them according to their position in this hierarchy, such a statement is fully justified. If we do not know why the size, shape, and number of the chromosomes of one species differ from those of another species, these chromosomal characters tell us no more than do differences in leaves, flowers, and fruits.

On the other hand, the evolutionist is not interested in classification as an end in itself, but as a means to the end of understanding the relationships between species, the processes which have brought about evolutionary diversification, and the directions which evolution has taken. If these are the goals which we set for ourselves, chromosomal differences have for us a meaning entirely different from morphological, physiological, and ecological differences. These latter differences represent the end products of long sequences of interaction between primary and secondary gene products, modified by the effects of the environment during development. Their causal significance in evolution is related only to the adaptive value which they possess and can confer on the evolving population.[163,209]

Chromosomal differences, on the other hand, reflect more or less directly the genic content of the individual. Difference in chromosome size may either reflect differences in the number of different kinds of gene products or proteins which the individual can produce, or reflect duplications of genes which influence the rate at which a particular kind of protein can be synthesized. As is pointed out in the next two chapters, differences in staining properties often, and perhaps usually, reflect differences in the timing of gene action. Differences in karyotype morphology reflect differences in gene arrangement which can affect drastically the way in which genes can become segregated and recombined in Mendelian heredity. Finally differences in chromosome number may reflect either differences in gene arrangement, or gene duplication, or both. In short, chromosomal differences reflect differences in the source of genetic variation, while morphological, physiological, and biochemical differences reflect differences in the products of gene action, modified by environmental influences. In order to understand evolution, we must become familiar with and take into account all of these differences. Moreover, we must also understand as well as possible the relationships between them. The remainder of this book will be devoted to exploring chromosomal differences, their relationships to other kinds of differences, and the significance of these relationships to our understanding of both the processes of evolution and the direction which it has taken.

2

Chromosomal Organization in Relation to Gene Action

The principal morphological features of chromosomes were outlined in the last chapter. The purpose of the present chapter is to analyse chromosomes more deeply, particularly in relation to their function of regulating gene action.

CHROMOSOMES AS ORGANELLES

At the outset, we must bear clearly in mind the differences between the chromosomes possessed by all organisms having true nuclei, the *eucaryote* organisms, and the 'chromosomes' of the *procaryotes,* which include only bacteria, blue-green algae, and viruses, if the last are to be regarded as organisms. The 'chromosome' of procaryotes consists only of a molecule of DNA, which may be temporarily associated with various proteins, but which otherwise lies in the centre of a cell not surrounded by an internal membrane. Furthermore, the bacterial chromosome does not appear to undergo any regular changes in conformation during the cellular division cycle.

On the other hand, chromosomes of eucaryotes, which are essentially similar in all organisms from *Amoeba, Chlamydomonas,* and *Euglena* to higher plants and animals, contain DNA complexed with proteins, RNA, and probably small amounts of lipids. These chromosomes during most of the cellular division cycle are enclosed in a nuclear membrane. Although less continuous than the cell membrane, the nuclear membrane can nevertheless selectively permit or exclude the entrance of various particles, including some protein molecules,[73] and can maintain a different physicochemical environment within the nucleus from that prevailing outside of it.[148] Moreover, chromosomes of eucaryotes undergo striking and regular changes in their outward appearance and their physicochemical state during the cellular division cycle. These changes are associated both with the orderly replication and distribution of the genes, and with the timing of gene action.

Each chromosome can, therefore, be best described as a *flexible organelle*. In order to understand the nature of chromosomes we must first examine their structure and their chemical composition. We must then look at the changes which they undergo during the mitotic and meiotic cycles. Finally, we must recognize the existence of special chromosomal changes associated with particular stages of development and differentiation of tissues.

THE STRUCTURAL ORGANIZATION OF CHROMOSOMES

The first problem which arises in connection with chromosome structure is the number of strands or double helices of DNA which each chromosome contains. Examination of chromosomes under the electron microscope, combined with biochemical studies, have shown that each strand of DNA, surrounded by the proteins with which it is complexed, forms a strand or **chromonema** which has a diameter about ten times that of the DNA double helix.[246]

The number of chromonemata of which chromosomes consist has been a subject of debate for almost half a century, and the matter is still by no means settled. The evidence from genetic segregation ratios and the expression of newly occurring mutations is most easily explained on the basis of a single DNA molecule per gene locus, except for the stages following interphase replication of DNA until the separation of daughter chromosomes at the beginning of mitotic anaphase.[59] Nevertheless, many years ago observers of untreated chromosomes under the light microscope saw good indications that chromosomes consist of many strands, and several more recent studies with refined techniques have reinforced this impression.

The most conspicuous landmark of the chromosome, the **centromere,** in large chromosomes treated with such reagents as colchicine and oxyquinoline, can be seen under the light microscope to be a compound structure.

Its nature has been explored in detail by Lima-de-Faria[144]. At late prophase, metaphase and anaphase of mitosis it is thinner and more weakly staining than the chromosome arms on either side of it. This condition is due to the fact that the condensation of the chromonemata which takes place in most regions during prophase affects the centromere less than other parts of the chromosome. Its compound nature consists of the presence of from two to five pairs of elliptical, more heavily staining

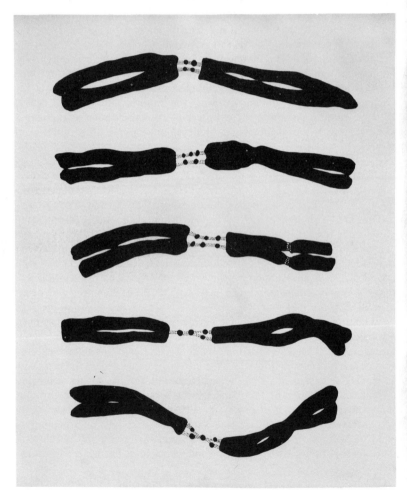

Fig. 2.1 Early metaphase chromosomes of root tips of *Hyacinthus orientalis*, showing chromosome structure of the centromeres. (From Lima-de-Faria.[144])

chromomeres, joined together by thin threads which stain more weakly, but which nevertheless contain DNA (Fig. 2.1). The chromomeres of the centromeric region resemble closely the similar structures which are found elsewhere along the chromosome at mid-prophase, and are described in a later section of this chapter.

The other distinctive region found in certain chromosomes, the terminal satellite, is associated with the accessory chromosomal organelle known as the nucleolus. At the prophase of both mitosis and meiosis the thread which connects the satellite with the end of the chromosome arm to which it is attached can be seen to pass through the nucleolus (Fig. 1.4). A variety of evidence, which will be reviewed in a later section of this chapter dealing with this body, indicates that the satellite connecting thread is a segment of the chromosomal DNA, known as the *nucleolar organizing region,* which codes for ribosomal RNA. It appears as a slender thread in late prophase and early metaphase chromosomes because much of the protein with which the DNA was complexed at earlier stages has been removed with the nucleolar material.

THE CHEMICAL COMPOSITION OF CHROMOSOMES

As can be seen in Table 2.1, the substance of which chromosomes consist, known as *chromatin,* contains more protein than DNA. The proteins found in chromatin are of several sorts. The most abundant are basic proteins, known as *histones,* which contain high amounts of the amino acids arginine, lysine, or both. Among the non-histone proteins, the most abundant are acidic proteins. At least some of these are similar to collagen, the protein of which animal connective tissue consists.[47,93] Other

Table 2.1   Chemical composition of chromatin from various sources. (Modified from J. Bonner et al.[19])

Source	% DNA	% Histone	Non-Histone Protein	% RNA	% of DNA active in transcription
Pea embryonic axis	39.0	40.0	11.0	10.0	12
Pea vegetative bud	40.0	52.0	4.0	4.0	6
Pea growing cotyledon	43.5	34.5	16.0	6.0	32
Rat liver	37.0	37.0	25.0	1.0	20
Cow thymus	40.2	46.0	13.5	0.3	15
Sea urchin blastula	39.0	41.0	19.0	1.0	10
Sea urchin larva (pluteus)	33.4	29.0	35.0	2.6	20

non-histone proteins present in chromosomes are the enzymes responsible for DNA replication and for the transcription of RNA. A noteworthy fact is that, while the histone content of chromosomes remains relatively constant during the life cycle of an organism, the content of non-histone proteins fluctuates widely, and this fluctuation is correlated with gene activity. As Table 2.1 shows, chromatin with a low content of non-histone protein has relatively few DNA sites which are active in the transcription of RNA. During the development of a particular organism, the number of active sites of DNA or gene loci increases proportionally to the increase in the amount of non-histone protein present.

In addition to DNA and proteins, chromatin contains small amounts of RNA. These amounts also fluctuate widely, and are roughly correlated with the amounts of non-histone protein. The function of this complex system of macromolecules has been the object of extensive research and controversy for many years, and the problem is by no means solved. A few facts are definitely established. When complexed with histone, DNA molecules are more condensed and tightly coiled than when the histone is removed. Furthermore, when histone is complexed artificially with DNA in a test tube, the ability of the complex to synthesize RNA by transcription is much less than the transcribing ability of naked DNA. The experiments of Bonner and others[19] indicate that natural chromatin also has a lower capacity for transcription than does naked DNA, but this conclusion has not been generally accepted. Furthermore, the messenger RNA's which are synthesized by DNA of chromatin derived from the cell nuclei of different animal tissues, such as bone marrow and thymus gland, contain some molecules which differ from each other in respect to their nucleotide sequences. This indicates that, although some of the genes which are actively transcribing RNA are the same for both tissues, others are active only in a particular tissue.[164] The existence of differential, selective gene repression on the part of the protein-RNA complex which is associated with DNA in chromatin is, therefore, reasonably well demonstrated.

CHROMOSOMAL VARIATION DURING THE MITOTIC CYCLE

The description of the chromosome as a flexible organelle is particularly appropriate in respect to the changes which chromosomes undergo during the mitotic cycle. On the basis of these changes, as well as of changes in the nuclear membrane, the mitotic cycle is ordinarily divided into the following six stages:

(1) *Metabolic stage,* between successive mitotic divisions;

(2) *Prophase,* during which the chromosomes contract or condense in preparation for mitosis;

(3) **Pro-metaphase,** a short stage lasting from the disappearance of the nuclear membrane, which ends the prophase stage, until the beginning of

(4) **Metaphase,** when the chromosomes are regularly aligned on the equator of the spindle;

(5) **Anaphase,** which includes the separation of the daughter chromosomes and their movement to the poles of the spindle, and ends with the formation of new nuclear membranes around the separated groups of daughter chromosomes;

(6) **Telophase,** during which the chromosomes become uncondensed and uncoiled, so that their identity is no longer recognizable under the microscope.

For the interval between mitoses, the term metabolic stage is far preferable to the term resting stage, which was used formerly by most cytologists. Chromosomes at this stage are far from being at rest. They are performing their most essential activities: transcription of RNA for protein synthesis and replication of DNA prior to the division of each chromosome into two daughter chromosomes.

During these successive stages, chromosomes undergo both physical and chemical changes. The physical changes consist partly of the coiling and folding of the chromonemata, which produces the shortening of the chromosomes from mid-prophase to metaphase, and of their uncoiling and unfolding during telophase. In addition, the two chromatids or half chromatids of which chromosomes consist become alternately closer together and more widely separated from each other. The correlation of these changes with the stages of mitosis is, however, not altogether clear.

The coiling and uncoiling of chromosomes during the mitotic cycle should not be confused with the coiled structure of the DNA molecules which they contain. The two types of coils are of a completely different order of magnitude. In DNA, each gyre has a width of 20 Å and the distance between successive gyres is 34 Å. Such dimensions are far below the resolving power of even the electron microscope. Their existence has been inferred from the appearance of DNA molecules when subjected to the X-ray diffraction technique.[244] The visible chromosomal coils, which were detected many years ago at metaphase in species with large chromosomes such as *Tradescantia* and *Trillium*,[41] are themselves of two dimensions. The gyres of the smaller coils have an amplitude of about 0.3 microns (3000 Å), which is more than a hundred times the amplitude of the DNA coil. Those of the larger coils, which are formed only at metaphase, have an amplitude of about 2.5 microns (25 000 Å).

The principal chemical event of the mitotic cycle is the replication of the DNA. This always occurs during the metabolic stage, but its timing relative to the end of telophase and the onset of prophase varies according to both the genetic constitution of the species and the intracellular environment, which in turn is determined to a considerable degree by the developmental stage of the tissue concerned.

The DNA replication itself does not occur simultaneously in all parts of the chromosome. In particular, those portions of chromosomes which stain differentially, and are known as heterochromatic regions, replicate their DNA at different times from the euchromatic regions. This subject will be taken up in more detail in the next section of this chapter. Some of the chromosomal proteins are synthesized in coordination with the synthesis of DNA, but other proteins are synthesized at various times during the mitotic cycle.[46]

The replication of DNA is only the first stage in the more complex process of chromosome division. As has been pointed out by Lima-de-Faria,[145] this is a summation of three processes: (1) reproduction, including DNA replication and protein synthesis; (2) individualization, in which two visibly separate strands, the future daughter chromosomes, become recognizable as chromatids; and (3) separation of chromatids to form the

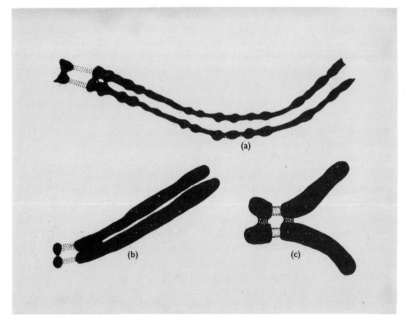

Fig. 2.2 Chromosomes at mitotic prophase in *Galtonia candicans*, ((a), (b), untreated) and at metaphase in *Hyacinthus orientalis* ((c), treated with colchicine), showing separation of centromere regions before that of the 'matrix' of proximal regions of the chromosome arms. (From Lima-de-Faria.[144])

daughter chromosomes. Reproduction occurs during the metabolic stage, as already stated; individualization occurs during early to middle prophase; separation begins in late prophase and continues through metaphase. In large plant chromosomes, different regions of the chromatids separate from each other in a regular succession (Fig. 2.2). The distal and median regions of the chromosome arms separate first, and are completely free from each other at late prophase, as can easily be seen in chromosomes treated with colchicine before fixation. The centromeres separate at pro-metaphase, and are completely distinct from each other throughout metaphase. Finally, the proximal regions of the chromosomes, those nearest to the centromere and on either side of it, separate at late metaphase and early anaphase.[145]

As is mentioned earlier in this chapter, much evidence suggests that at least in plants with large chromosomes, each chromosome consists of two or four chromonemata. If this is the case, then reproduction takes place one or two division cycles ahead of separation.

The mechanism of separation involves chiefly the formation of the mitotic spindle, the regular orientation of the chromosomes at its equator, and the separation of the daughter chromosomes at anaphase in response to the activity of the spindle fibres. The physicochemical factors involved in this mechanism are enormously complex and poorly understood. Moreover, they appear to be very similar in all eucaryote organisms, and complete knowledge of them is consequently not essential to a comparative study of chromosomal variation.

HETEROCHROMATIZATION AND ITS SIGNIFICANCE

The most obvious linear differentiation of chromosomes, particularly at prophase, is into regions which stain darkly with the usual stains, and those which stain more lightly, and so are less conspicuous. In the terminology originally proposed by Emil Heitz, the dark regions are called *hetero-chromatic* and the light regions *euchromatic*. Figure 1.3 shows a prophase nucleus of *Plantago insularis*, in which these regions are clearly evident.

Chromomeres, chromocentres and knobs

In chromosomes, at prophase as well as the metabolic stages, three kinds of heterochromatic regions are most often found. These are *chromomeres, chromocentres,* and *knobs.* Chromomeres are regular features of all prophase chromosomes which are large enough to reveal them. Furthermore, as shown many years ago by John Belling, the number, size distribution, and pattern of arrangement of the chromomeres are specific

Fig. 2.3 The three chromosomes of *Ornithogalum virens* at meiotic prophase (pachytene, top row) and at second microspore prophase (two cells, middle and bottom rows). Note that the distance between chromomeres is about the same at the two stages, but that the number of chromomeres per chromosome is much less in the mitotic prophase than at meiosis. (From Lima-de-Faria *et al.*[147])

for any particular species at any one stage of development. On the other hand, the number of chromomeres per chromosome varies from one stage of development to another. In particular, it is several times as large in the long, very slender chromosomes of meiotic prophase (pachytene) as in the much shorter chromosomes of mitotic prophase. This difference is clearly seen in species such as *Ornithogalum virens* (Fig. 2.3), in which the chromosomes are relatively large and few in number ($n = 3$).[147] In fact, careful studies of chromomere distribution at different stages in a number of species have shown that the distance between adjacent chromomeres is remarkably constant both in the same species at different stages and in different species having chromomeres of very different sizes (Table 2.2). This fact supports the conclusion which a number of cytologists have reached that the formation of chromomeres is based upon some general

Table 2.2   Average chrommere distance in different species. (From A. Lima-de-Faria et al.[147])

Family	Species	Stage	Average chromomere distance in microns
Liliaceae	Ornithogalum virens	Pachytene	1.07
	Ornithogalum virens	Second microspore prophase	0.89
Liliaceae	Agapanthus umbellatus	Pachytene	0.84
	Agapanthus umbellatus	Prophase II	0.95
	Agapanthus umbellatus	Prophase root tips	1.09
Gramineae	Secale cereale	Pachytene	0.82
Solanaceae	Solanum lycopersicum	Pachytene	0.83
	Solanum lycopersicum	Prophase root tips	0.96
Labiatae	Salvia horminum	Pachytene	1.06
	Salvia horminum	Second microspore prophase	0.94
		Mean	0.95

property of chromatin, and is probably related to the coiling of the chromonemata.

Knobs are spherical heterochromatic bodies which range in size continuously from those hardly wider than chromosomes to structures

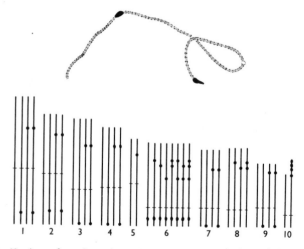

Fig. 2.4 Knobs of maize chromosomes, as seen at pachytene. Above, Chromosome 3, showing one terminal and one interstitial knob. Below, diagrams of the ten chromosomes, showing various positions of knobs found on each chromosome in different varieties. (From Longley.[149])

many times the diameter of the chromosome. They are best known in maize and its relatives (Fig. 2.4), but they are wide-spread in other plants. They may occur either on the ends or in the middle of chromosomes, as in *Ornithogalum* (Fig. 2.3). Within any genotype, their number and position are constant, but their size may vary greatly from one developmental stage to another (Fig. 2.3). Knobs are particularly valuable chromosomal markers, both for distinguishing between the chromosomes of related species and races, and for recognizing the presence of segmental interchanges, both natural and induced.

Chromocentres are heterochromatic regions of varying size which in many species occur near the centromere, in proximal regions of chromosome arms. They are most commonly recognized in the metabolic stage and in early prophase. At mid-prophase of both mitosis and meiosis, many chromocentres become resolved into strings of chromomeres which are larger in size than the chromomeres found in the more distal regions of the chromosome, and which gradually diminish in size from the more proximal to the more distal parts of the centromeric region (Fig. 2.5). In some metabolic nuclei, particularly those of the salivary glands in Diptera, the

Fig. 2.5 Homologous chromosomes of *Agapanthus umbellatus* at meiotic prophase (pachytene, top), middle prophase II of meiosis, and middle prophase of mitosis (bottom), showing chromomere gradient. (From Lima-de-Faria.[143])

chromocentres of different chromosomes fuse to form a single large chromocentre. As in the case of knobs, the appearance of chromocentres varies greatly according to both the stage of the mitotic cycle, and the developmental stage of the organism concerned. Nevertheless, they are constant within a species, or a group of related species, when studied at a particular stage of development, and exhibit systematic variations from one species or species group to another. Hence comparative studies of the distribution of chromocentres and similar concentrations of heterochromatic regions near the centromere have considerable evolutionary value. This value may increase greatly when we know more about the connections between visible heterochromatization and chromosome chemistry as well as gene action.

Heterochromatization of sex chromosomes

In addition to these heterochromatic regions which exist in many chromosomes at various positions along their length, some heterochromatic elements may include entire chromosomes or large portions of them. In plants, the commonest heterochromatic chromosomes are the small accessory or B-type chromosomes which occur in varying numbers in many species. These will be discussed in the next chapter. In many species of animals, one of the two sex chromosomes is entirely heterochromatic, and the other usually contains large heterochromatic regions. Sex chromosomes occur much less commonly in plants than in animals, but in many dioecious plant species there is completely visible chromosomal differentiation between the sexes. Where such differentiation occurs, however, as in *Lychnis* (*Melandrium*), *Rumex,* and many species of liverworts and mosses, it may involve partial or complete heterochromatization of one or both sex chromosomes, as in animals. This subject will be discussed on a comparative basis in the next chapter.

Differential gene replication and activation in heterochromatic regions

One well established chemical fact about heterochromatic regions is that their DNA is replicated at a different time from that of euchromatic regions. Most heterochromatic regions have late replicating DNA,[146] but in the root tips of the orchid *Spiranthes* and some other plants, early replication of DNA in the heterochromatic regions has been found.[226] The reason for these differences is not clear.

Another fact which is now supported by a variety of evidence is that genes are inactive when chromatin is in the condensed or heterochromatic condition.[25] This inactivity of heterochromatic regions led cytogeneticists many years ago to suspect that these regions did not contain genes. Now, however, genes have been localized in both *Drosophila* and the tomato in regions which are heterochromatic at most stages of development. Moreover, the recognition that genes consist of DNA, and that this substance has actually a higher concentration in heterochromatic than in euchromatic regions leads to the conclusion that if they differ at all in this respect, heterochromatic regions must actually contain more genes than euchromatic regions of the same size. Coupled with this conclusion, however, is the fact which has also been recently demonstrated with great clarity, that in multicellular animals and plants many genes are active only for a relatively short period during the development of the organism.

Sex chromosomes of mammals

The evidence which has associated heterochromatization with gene inactivation, and euchromatization of normally heterochromatic regions with stage-specific gene activity has come principally from animals. Two lines of evidence are particularly convincing. One is from the phenomenon of facultative heterochromatization, in which two homologous chromosomes having nearly the same gene content exist side by side in the same cell, one of them heterochromatic and the other euchromatic. This situation is most strikingly exemplified by the two X chromosomes of a female mammal. Early in embryonic development one of the X chromosomes becomes heterochromatic and the other remains euchromatic. The association of this change with inactivation of genes is clear in the case of animals heterozygous for genes located on the X, and therefore sex-linked. An example is the tortoiseshell cat. This colour pattern exists only in females. Males, which have only one X, are either black (B) or yellow (b). In females having the constitution Bb, either the X chromosome containing the B allele becomes heterochromatic or that containing b becomes heterochromatic and genetically inactive. In this condition, it becomes visible in metabolic nuclei as a deeply staining body. The allele in the chromosome which remains euchromatic and active determines the colour of the hair cells which contain it: black if it contains the allele B and yellow if it contains b. Since in the tortoiseshell cat black and yellow hairs are distributed over the body in a more or less random fashion, we can conclude that, in the embryonic cells which will later give rise to hair cells, either the B-containing or the b-containing X chromosome is heterochromatized and inactivated, more or less at random.

The Y chromosome of Drosophila

A second line of evidence indicating that chromosomes in the heterochromatic condition are inactive, and that euchromatization brings about activity, has been obtained by studies of the Y chromosome in species of *Drosophila*.[103] This chromosome is entirely heterochromatic, and contains no genes which can be detected by applying to mutations the usual linkage and crossover tests. Geneticists have known for many years, however, that males which lack this chromosome, although they are vigorous and completely normal in appearance, are nevertheless sterile. The job which the Y chromosome performs in rendering the males fertile was discovered by Oswald Hess and G. F. Meyer in an intensive comparative study of the structure and development of spermatocyte nuclei in Y-bearing fertile flies and in sterile flies which lack either an entire Y chromosome or part of it.

The developing spermatocytes of normal, Y-bearing males contain an elaborate system of cytoplasmic organelles, which are apparently essential for their development. In males lacking a Y chromosome, these organelles are absent, and the sperm does not develop. Moreover, if portions of the Y chromosome are removed by X-ray breakage, certain of the spermatocyte organelles fail to form, and the development of the spermatocytes is disturbed to varying degrees. This demonstration that the Y chromosomes are genetically active in developing spermatocytes but inactive in all other kinds of cells and tissues was accompanied by cytological observations which showed that in spermatocytes portions of the Y chromosome become euchromatic.

Although the relationship between heterochromatization and gene action is still unclear and imperfectly known, two conclusions can be reached even at the present time. First, intensive studies of heterochromatization in relation to gene action in genetically well known organisms are among the most fruitful approaches to an understanding of gene action in higher organisms in general.[22] Second, comparative studies of the patterns of heterochromatization of chromosomes in related species will eventually provide much insight into the kinds of mutational changes which have been responsible for the origin of new patterns of gene action in development, and hence for new organizations of adult structures. The progress which has been made in this comparative field will be reviewed in the next chapter.

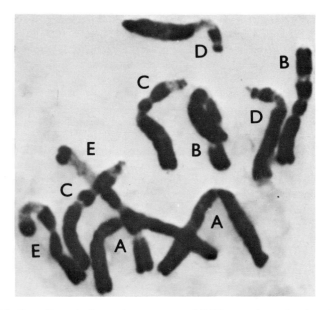

Fig. 2.6 Somatic metaphase chromosomes of *Trillium ovatum*, showing allocyclic regions which are revealed by cold treatment. (Photograph by I. Fukuda.)

ALLOCYCLY AND ITS SIGNIFICANCE

A phenomenon related to heterochromatization, which can be regarded as persistent chromosomal condensation, is the decondensation or relaxation of certain chromosomal regions under the influence of certain artificial treatments (Fig. 2.6), the most effective of which is exposure of living chromosomes to near freezing temperatures for several days.[95] This phenomenon is termed *allocycly*, indicating that the coiling cycle of the regions which react to these treatments differs from that of the rest of the chromosome. An alternative term, nucleic acid starvation, has been used in the past, but is inappropriate, since chemical analyses indicate that no alteration of DNA metabolism is involved.[248]

Comparative studies of allocyclic regions in *Trillium* and other plant genera show that many populations are polymorphic for chromosomes having different patterns of allocycly, and that such patterns have regular geographical and ecological distributions. They will be discussed in the next chapter. Unfortunately, nothing is known about the relationship, if any, between allocycly and gene action.

THE ROLE OF THE NUCLEOLUS

One of the most conspicuous organelles in the nucleus is the *nucleolus*. Nucleoli are found in most eucaryote nuclei, though their number and size vary both from one kind of organism to another, and from one stage to another in the development of a single organism. They are formed at telophase, persist throughout the metabolic stage, and usually disappear in middle or late prophase. They consist chiefly of an RNA–protein complex.[16]

A distinctive feature of nucleoli is their attachment at certain stages of the mitotic cycle, particularly early prophase, to a restricted region on one or more chromosomes of the complement (Fig. 1.4). This region is known as the *nucleolar organizer region*. It appears in late prophase and metaphase as a slender thread which connects a satellite with the rest of the chromosome. Both genetic tests and experiments using tritiated uridine indicate that at least the RNA contained in the nucleolus is transcribed from genes located in this region.[16]

The function of the nucleolus, long unknown, has now been revealed in its general features by a combination of genetic, cytological, and biochemical investigations. A mutant of the clawed frog, *Xenopus*, which lacks a nucleolus, is lethal in the homozygous condition, indicating that this organelle is essential for normal development.[24] The development of these homozygotes is blocked in the late blastula stage of the embryo at a time when many ribosomes are formed in normal embryos, and when a great increase in the synthesis of messenger RNA takes place. The heterozygous embryos develop into larvae and adults, but at the critical developmental stage they are deficient in ribosomes. This evidence suggests that nucleoli are largely responsible for the synthesis of ribosomes.

Even stronger evidence for this theory has been obtained by the research of F. M. Ritossa and his associates[186,187] on the bobbed locus of *Drosophila*. Mutations at this locus, which produce flies with much shortened bristles, have been known since the earliest days of *Drosophila* genetics. The locus is on the sex chromosomes, and is unique in that a normal 'allele' of bobbed exists on the Y as well as on the X chromosome. Mapping techniques indicate that bobbed is at or near the nucleolus organizer region. Ritossa has shown by DNA hybridization techniques that bobbed flies have less ribosomal RNA than normal flies, and that they are also partly deficient in DNA which codes for ribosomal RNA. This evidence indicates that bobbed 'mutations' are actually deficiencies for part of the nucleolar organizer region, and that the DNA of this region codes for ribosomal RNA. Since this kind of RNA exists in high concentrations in the nucleoli themselves, and ribosomes are first contained in nucleoli and then released into the cytoplasm,[179] one may conclude that these organelles are sites for the synthesis of ribonuclear proteins, and for the formation of ribosomes. When released into either the nuclear sap or the cytoplasm, ribosomes become complexed with messenger RNA, and participate in protein synthesis.

One might justly ask the question: Since the nature and function of ribosomes in eucaryotes is the same as in bacteria and blue-green algae, which lack nucleoli, why should ribosome synthesis in eucaryotes take place in these specialized organelles? The answer lies probably in the connection between the concentration of ribosomes and the rate of protein synthesis. Research with bacteria has shown that for any polyribosomal messenger RNA complex protein synthesis by translation occurs at a constant rate, regardless of the environment. The rate of protein synthesis can be increased only by increasing the number of polyribosomal messenger RNA complexes which are simultaneously engaged in synthesis. We might suspect that in the relatively undifferentiated cells of bacteria, wide fluctuations in the rate of protein synthesis have no particular adaptive value, but that the much more complex development of the highly differentiated cells of eucaryotes periodically requires greatly increased rates of protein synthesis. Nucleoli, because of their ability to enlarge and multiply during these periods, serve as amplifiers of this process.

This hypothesis is well supported by the changes which nucleoli undergo at various stages of development in both animals[43,80] and plants. Large cells, such as trichome initials and sporocytes, usually have very large nucleoli during the period when they are actively growing and presumably are

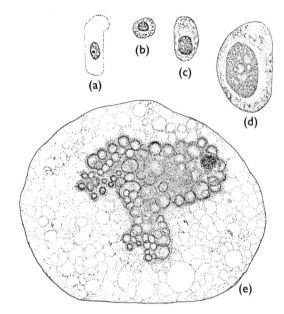

Fig. 2.7 Drawings showing the growth of the nucleus and nucleolus during the development of the very large cell of *Acetabularia Wettsteinii*. **(a)–(d)** Young plants at ages of 5, 24, 39 and 73 days. **(e)** Nucleus of a full grown cell, 5 months old, showing the vesiculate nucleolus. (From Schulze.[192])

synthesizing large amounts of protein. This condition is most conspicuous in the alga *Acetabularia*, which during its entire period of growth up to a length of about 5 cm consists of a single gigantic cell. Early in the growth of this cell, its nucleus becomes greatly enlarged, and encloses an enormous nucleolus, which may be as much as 100 microns in diameter (Fig. 2.7). In higher plants, the nuclei of the apical meristem during the transition from the vegetative to the reproductive state contain very large nucleoli.[130] At this time, large amounts of protein are being synthesized.

STAGE-SPECIFIC ENDOPOLYPLOIDY AND POLYTENY

A regular feature of the development of some plant as well as animal tissues is the formation of giant nuclei containing many times the usual somatic number of chromosomes. These occur in the large epidermal cells of certain monocotyledons, such as Gramineae and Araceae; in the large, water containing cells in the interior of the leaves of some succulents; in epidermal hairs or trichomes, both unicellular and multicellular; in the tapetal cells of anthers, surrounding the sporocytes; in antipodal cells of the embryo sac, particularly those of grasses and some Ranunculaceae; and in the haustorial suspensor of the embryo in Leguminosae, Scrophulariaceae, and other families.[173,232,233] In the developing multicellular trichomes of such genera as *Saponaria* and *Cucurbita*, mitoses with polyploid chromosome numbers of increasingly high levels can be found (Fig. 2.8). The large nuclei which result from this polyploidization usually have much enlarged nucleoli.

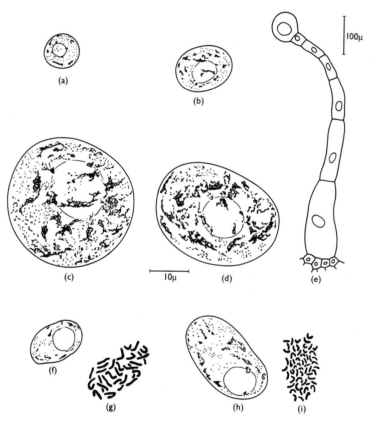

Fig. 2.8 Endopolyploid nuclei in *Saponaria ocymoides*. **(a)** Diploid nucleus from epidermis of calyx. **(b)** Diploid and **(c)**–**(d)**, 16-ploid nuclei from cells of trichome (shown at right, **(e)**). **(f)** and **(g)**, Diploid; **(h)** and **(i)**, tetraploid nuclei from pith of the stem; at metabolic stage and mitotic metaphase respectively. (From Tschermak-Woess and Hasitschka.[233])

The significance of these large, polyploid nuclei is obscure. They look so much alike in cells from very different tissues that one can hardly relate them to specific processes of differentiation. Nevertheless, their regular occurrence in certain tissues indicates that they have a functional significance. By analogy with the development of nucleoli in large cells which synthesize large amounts of protein, and in accordance with the fact that the polyploid cells themselves usually contain very large nucleoli, the hypothesis may be suggested that the polyploid condition of these cells also helps them to synthesize large amounts of protein in a relatively short period of time. Since both gene loci capable of transcribing messenger RNA as well as the nucleolar organizing regions which transcribe ribosomal RNA are multiplied many times, these cells should be capable of forming a much larger number of polyribosomal messenger complexes than could diploid cells.

THE BASIS OF CHROMOSOME PAIRING AT MEIOSIS

The chromosomal changes which were discussed in previous sections of this chapter were all associated with the function of chromosomes in regulating gene action. This final section will take up the most important chromosomal function related to gene recombinations: the pairing of chromosomes at meiosis prior to their reduction in number.

The course of meiosis

The numerous differences between the meiotic divisions in the anthers or ovules and ordinary mitosis in somatic cells involve two kinds of events: (1) The much longer prophase before the dissolution of the nuclear membrane, during which homologous chromosomes pair and exchange segments; (2) the different way in which the chromosomes segregate at the first and second divisions.

Cytologists have customarily recognized five subdivisions of the meiotic prophase: *leptotene, zygotene, pachytene, diplotene,* and *diakinesis.* With reference to the most important event of meiosis, the visible pairing of homologous chromosomes, they may be arranged into three groups, as follows: before pairing, leptotene; during pairing and cytological crossing over, zygotene and pachytene; contraction after pairing, diplotene and diakinesis.

The nature of chromosome pairing

In higher plants, the close pairing of homologous chromosomes at zygotene begins at various points and continues, zipper fashion, until the entire length of each chromosome is paired. At pachytene, the homologues are so closely paired that in many cells the thick threads which they form cannot be recognized as double throughout their length (Fig. 2.12). Each chromosome consists of an alternation of bead-like heavily staining condensed portions known as *chromomeres* which are connected to each other by more weakly staining threads.

Electron micrographs of both animal and plant[235] cells at pachytene have revealed the presence of *synaptinemal complexes* involving the paired chromosomes. These complexes consist of two lateral elements which are electron dense, and apparently represent homologous chromatids. Between them is an electron-dense central element, flanked by two clear regions, which in particularly clear pictures can be seen to be traversed by fine threads. Although the significance of synaptinemal complexes is not yet clear, their presence at the critical pairing stages of meiosis in all eucaryotic organisms investigated strongly suggests that they are the structures in which the close association of homologous chromosomes necessary for crossing over takes place.

To a cytologist interested in chromosomes and evolution, the great significance of observations at pachytene lies in the fact that in plants, as well as in all animals except for dipterous flies, the precise resemblances or differences between homologous chromosomes can be seen only when they are closely paired at this stage. For this reason, cytogenetic analyses of species and hybrids in genera which have easily observable pachytene stages are particularly desirable. Ease of observation depends upon three factors: (1) a large nucleus relative to the size of the chromosomes, so that the latter can be easily spread out; (2) a relatively small number of chromosomes; (3) distinctive patterns of chromomeres or the distribution of heterochromatic regions, so that the chromosomes can easily be distinguished from each other.

Crossing over, chiasma formation, and the appearance of bivalents

The most conspicuous feature of late prophase and metaphase of meiosis is the formation and appearance of *bivalents,* made up of two homologous chromosomes joined together. After pachytene, homologous chromatids tend to separate from each other, so that the chromosomes forming the bivalent are closely joined only where chromatids belonging to homologous chromosomes exchange partners. The visible evidence of this exchange is a conspicuous node or cross, known as a *chiasma.* The morphological appearance of the bivalents at late prophase and metaphase depends in large part upon the number and location of chiasmata, as well as the location of the centromere, whether median, subterminal, or terminal (Fig. 2.9).

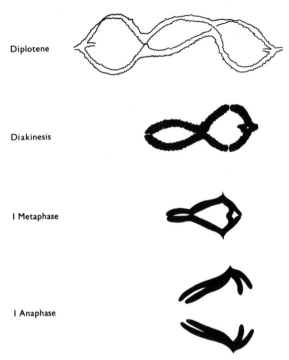

Diplotene

Diakinesis

I Metaphase

I Anaphase

Fig. 2.9 A typical bivalent at late prophase, metaphase, and anaphase, showing the changes in appearance caused by contraction and alteration of chiasmata.

The clearest evidence to show that chiasmata as seen at diakinesis and metaphase represent physical crossing over between homologous chromatids at an earlier stage is obtained in plants with large chromosomes, numerous chiasmata, and little movement of chiasmata along the chromosome during prophase. If in a plant having these characteristics the homologous chromosomes can be differentiated from each other by means of cytological markers, in the form of differential terminal knobs, the cytologist can tell whether the exchange of chromatids which he sees in a chiasma represents exchange of pairing between unbroken chromatids, which are closely associated alternately with sister chromatids and with those belonging to homologous chromosomes, or whether close association is always between sister chromatids. In the latter case, those chromatids which participate in chiasmata have acquired, by crossing over, a constitution made up partly of genes derived from one homologue, and partly derived from the other (Fig. 2.10). Numerous cytological observations, like the one illustrated in Figure 2.10, indicate that the presence of

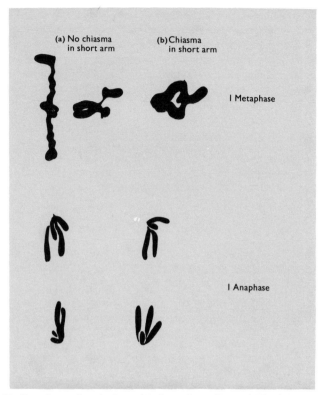

Fig. 2.10 Drawings of metaphase bivalents in a clone of *Lilium* heterozygous for an extension of the short arm in one of the homologues. **(a)** Without previous crossing over and chiasma formation in the short arm, so that separation at first anaphase (below) is reductional, the short chromatids going to one pole and the long chromatids to the other. **(b)** After crossing over and chiasma formation in the short arm, so that first anaphase separation is equational, one long and one short chromatid going to either pole. (From Brown and Zohary.[26])

chiasmata indicates surely that chromatids have become associated in such a way that the chromatid which can be seen to 'cross over' from one homologue to the other actually does contain a combination of genes derived from both of the parental homologues.

Careful observations of successive stages of meiotic prophase in favourable organisms have shown that chiasmata may either retain the same position along the chromosome which they occupied when formed by crossing over, or they may become **terminalized** to various degrees. Terminalization is the movement of the chiasma along the chromosome, distally from the point where crossing over occurred. In many plants, such as *Oenothera*

(Fig. 4.5), the chiasmata may become so completely terminalized at meta-phase that only the ends of the chromosomes are associated with each other. This is particularly true of plants having small chromosomes, but is found in many plants with large chromosomes, such as *Tradescantia* and the American species of *Paeonia*. The relationship of terminalization to the fertility and success of interchange heterozygotes is discussed in Chapter 4.

Factors which affect chromosome pairing

The most important and generally effective factor which determines the nature of chromosome pairing is the structural and chemical similarity known as homology. The clearest evidence of its effectiveness is the

Fig. 2.11 Meiotic prophase (diakinesis) in a sporocyte of *Ophioglossum reticulatum,* showing about 630 bivalents. (From Ninan.[174])

fidelity with which each paternal chromosome associates with its counterpart among the maternal chromosomes, thus making Mendelian segregation of genes possible. In many plants having high numbers of chromosomes, such as the adder's tongue fern (*Ophioglossum*, Fig. 2.11), the regular formation of hundreds of bivalents in every meiotic cell, each of which represents the association of a chromosome with only one specific mate out of the hundreds which surround it, is nothing short of miraculous.

The degree of homology between chromosomes depends at least in part upon their similarity with respect to gene arrangement. In a plant having two sets of chromosomes that contain corresponding or allelic gene loci arranged in identical order, the close pairing of homologous chromosomes can be observed at pachytene in favourable material (Fig. 2.12). If

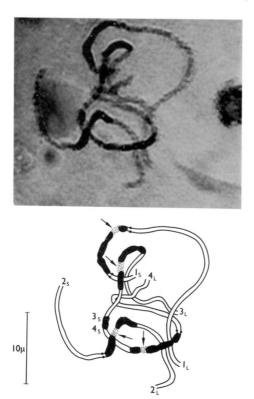

Fig. 2.12 Mid prophase of meiosis (pachytene) in *Plantago insularis* (*n* = 4), showing intimate pairing of the homologous chromosomes. The long and short ends of the chromosomes are indicated by the symbols 1L–1S, etc,; centromeres by arrows. (Photo by Alva Day Whittingham.)

otherwise homologous chromosomes differ with respect to inverted or translocated segments which are of considerable size, most of the allelic loci still pair closely with each other. This pairing gives rise to the familiar loop configurations in inversion heterozygotes (Fig. 4.2) and cross configurations in translocation heterozygotes (Fig. 4.4). Nevertheless, the presence of such structural heterozygosity does reduce the amount of association between allelic loci. When inversion loops are formed, the chromosomal regions in the vicinity of the loop are less closely associated than elsewhere along the chromosome. In the case of translocation configurations, the point of exchange between chromatids belonging to different chromosomes may be some distance away from that where the translocated segment is attached, so that considerable segments of chromosomes are paired in a non-homologous fashion. If inversions or translocations are small, the characteristic loop or cross configurations may not be formed at all, so that the rearranged regions are paired in non-homologous fashion.

Preferential pairing in polyploids

The dependence of pairing upon structural, and probably also chemical, similarity between chromosomes is strongly supported by the effects on pairing of doubling the chromosome number. In a hybrid between two widely different species, such as radish (*Raphanus sativus*) and cabbage (*Brassica oleracea*), the parental chromosomes are so dissimilar that little pairing occurs at meiotic prophase, and at metaphase most of the chromosomes are present as univalents. If, however, the chromosome number of the hybrid is doubled, so that each cell contains two chromosomes of each kind that are exactly alike, a full complement of closely paired bivalents is formed. Still more significant are the effects of doubling a hybrid between parents having chromosomes sufficiently alike so that they can pair, but different enough so that the amount of pairing in the undoubled hybrid is reduced, or the chiasma frequency is lowered. When such hybrids are doubled, as exemplified by *Primula kewensis* (Fig. 5.5), pairing is largely confined to chromosomes derived from the same parent, which are exactly alike because they were produced by replication of the same original chromosome. Partly homologous, or **homoeologous,** chromosomes may pair with each other if two completely homologous chromosomes are not present in the same cell. If, however, a cell contains four sets of chromosomes, two of which are partly homologous and two fully homologous with each other, the fully homologous sets exhibit **preferential pairing** with each other. They may form exclusively bivalents, so that pairing between homoeologous chromosomes is completely eliminated, or they may form varying numbers of multivalent configurations, which involve pairing between both fully homologous and homoeologous chromosomes. The significance of this phenomenon is discussed further in Chapter 5.

Genes which affect chromosome pairing

Two groups of factors affect chromosome pairing: the similarity between the chromosomes themselves and the cellular environment which prevails at meiosis. The latter can be affected either by drastic alterations of the external environment, such as cold or heat shocks, or by genes which presumably control the numerous enzymatic reactions required for meiosis. Only the latter will concern us here.

In respect to their evolutionary significance, two kinds of genic effects must be recognized: those which affect pairing indiscriminately, and those which intensify preferential pairing. Individual recessive genes exist which in the homozygous condition cause the chromosomes to be unpaired at meiotic metaphase, either through the failure of the initial association at zygotene and pachytene, or through the failure of chiasma formation. Such genes are relatively uncommon, and in natural populations would be quickly eliminated or greatly reduced in frequency, because of their harmful effects on fertility.

From the standpoint of plant evolution, the gene located on chromosome 5B of wheat is of the greatest significance, and has deservedly been subjected to intensive study.[185] When this chromosome is present, the 42 chromosomes of bread wheat regularly form 21 bivalents, the chromosomes of which are closely paired, each bivalent normally containing two or three chiasmata. If, however, this chromosome is lacking, the plants form a variable number of trivalents or quadrivalents, and meiosis is much more irregular. This is because bread wheat is a segmental allopolyploid (see p. 130). Its gametic set of 21 chromosomes consists of three sets of seven. Each of the chromosomes of a set is partly homologous to and able to pair loosely with a corresponding or homoeologous chromosome in one or both of the other two sets. Pairing between these homoeologous chromosomes, combined with pairing between the completely homologous chromosomes which the plant has received from opposite parents gives rise to the observed multivalents. The 5B gene, therefore, has the specific effect of eliminating pairing between partly homologous chromosomes without affecting at all the pairing between completely homologous chromosomes.

The evolutionary significance of this gene is twofold. First, by discriminating between homoeologous and completely homologous pairing, it demonstrates the importance of specific features of chromosome structure and degree of evolutionary divergence as determinants of pairing in the absence of other controlling factors. Second, it shows that polyploids which have resulted from chromosome doubling in hybrids between rather closely related species can evolve regular meiosis by selection for specific genes or gene mutations rather than by the much more complex process of selection for chromosomal rearrangements. This topic is discussed further in Chapter 5.

Chromosome pairing as a measure of evolutionary relationship

The question about chromosome pairing which most concerns the evolutionist is the following one: To what extent does the closeness of pairing reflect the degree of genetic relationship between two plants? The answer to this question is not simple. On the one hand, the complex series of interacting factors that determine pairing can vary in so many different ways that no quantitative index can be devised which will establish an absolute relationship between the degree of pairing of the parental chromosomes in an F_1 hybrid and the closeness of relationship between its parents. On the other hand, in many groups of plants the correlation between pairing in a hybrid and the degree of relationship between its parents is good enough so that we cannot disregard it altogether.

The best way out of this difficulty is to be fully aware of the factors, both chromosomal and genic, which can influence pairing, and to evaluate them as carefully as possible in each separate instance. Two generalizations hold well enough so that they can be used as guide lines for such evaluation. The first of these is that the genes which affect chromosome pairing by their influence on the cellular environment, including protein synthesis, do not affect differentially the association of individual chromosome pairs. Their effects are general, and if under their influence there is variable chromosome pairing, this variability exists between cells as well as within cells. Secondly, chiasma formation is proportional to chromosome length, with some qualification.[41] If, therefore, failure of metaphase pairing is due to genetically or environmentally controlled effects of chiasma formation, the smaller chromosomes of a set will be more strongly affected than the larger ones.

These principles are best applied to analyses of the relative amount of homology between different parental chromosomes which have been brought together in a hybrid. If one group of chromosomes derived from opposite parents consistently forms more closely associated bivalents than another, close genetic similarity between these pairing chromosomes can be postulated. On the other hand, when two different hybrids are compared, a lower amount of pairing of all chromosomes in one of them may be due to the action of one or a few individual genes which interfere with overall pairing, and thus may be unrelated to the actual genetic similarity of the parental chromosomes themselves. Analyses of chromosome pairing can provide valuable indications of relationships between species if they are used carefully, and with full recognition of all of the factors involved.

3

Variations in Size and Organization of the Chromosomes

In the last two chapters, the importance of chromosomes in the regulation of gene action was pointed out, and some of the changes related to this function which they undergo during the life cycle of an individual organism were described. These changes are of evolutionary significance because different kinds of plants have very different life cycles, and diverse adaptations to their environment during the various stages of their life cycle. Hence we would expect to find genetically controlled chromosomal differences between organisms parallel to those existing at different stages in the same organism. This is exactly what we find, particularly in relation to differential condensation or heterochromatization, as well as to the development of nucleoli. In addition, the size of the chromosomes and the total DNA content of the nucleus vary greatly from one group of plants to another. At least some of this size variation is also probably related to gene action in development, although the nature of this relationship is still obscure. The objective of the present chapter is to review the available information about chromosome sizes and degrees of condensation on a comparative basis, and to suggest tentative working hypotheses about the meaning of these data.

VARIATIONS IN CHROMOSOME SIZE

The total mass of the chromosomes in a nucleus is closely correlated with its DNA content. This correlation has been established on the basis of parallel measurements of chromosome volume and DNA content in about 200 species.[8] It suggests that at any particular stage of development the ratio of DNA to protein is relatively constant in eucaryote organisms. Because of this correlation, we can make reasonable inferences about the DNA content of an organism, and hence of the amount of genic material which it possesses, by examining its chromosomes. Such comparisons must be made at comparable developmental stages and under similar environmental conditions.

Table 3.1   DNA content of nuclei in various animals and plants. Content recorded as per cent of that found in *Lilium*. (Data from Sinsheimer[195] and from Sparrow and Evans.[202])

Microorganisms	Content	Plants	Content
T2 Bacteriophage	0.0004	*Chlorella*	0.25
Escherichia coli	0.02	*Scenedesmus*	0.9
Animals			
Sponge	0.12	*Arabidopsis*	4.0
Coelenterate	0.6	*Glycine max*	6.5
Echinoderm	1.8	*Vicia faba*	38.4
Lungfish	100.0	*Zea mays*	14.1
Teleost fishes	1.0–3.0	*Allium cepa*	54.3
Salamander (*Amphiuma*)	168.0	*Tradescantia paludosa*	59.4
Frog	15.0	*Lilium Henryi*	100.0
Turtle	5.0		
Mammals (man, rat, cow)	5.8–6.4		

Table 3.1 gives the nuclear DNA content of a representative series of plants, and also values for a species of bacterium (*Escherichia coli*). From these data we see that the lowest value recorded for a eucaryote plant, that in the unicellular green alga *Chlorella*, is more than 12 times that in the bacterium, and in turn is only 1/16 that found in the flowering plant *Arabidopsis*, the lowest value recorded for flowering plants. Values for other algae and for bryophytes are for the most part intermediate between those for *Chlorella* and for *Arabidopsis*.

Even more striking, however, are the differences in DNA content between different species of angiosperms. The nuclei of *Lilium* contain 25 times as much DNA as those of *Arabidopsis*, a difference considerably greater than that between *Arabidopsis* and *Chlorella*. While the value for *Lilium* is certainly one of the highest in the plant kingdom, it is nearly equalled by that for *Trillium*, and is probably exceeded by species of the mistletoe genus *Phoradendron* (Loranthaceae) which have 28 chromosomes even larger than the 24 found in *Lilium*.[243] The fact must be mentioned that these four genera of angiosperms, as well as many others having either extremely high or extremely low amounts of DNA in their nuclei, are relatively specialized end points of their particular lines of evolution.

The reasons for these differences in DNA content and chromosome size are not well understood. Later in this chapter, four possible hypotheses to account for them will be discussed. First, however, we must review certain regularities in their distribution.

Regularities in variation

The first regularity is a progressive increase in chromosome size and nuclear DNA content as we go from viruses to bacteria, and then to algae and fungi, bryophytes, and vascular plants. To be sure, there are great variations between different genera and species in certain of the lower

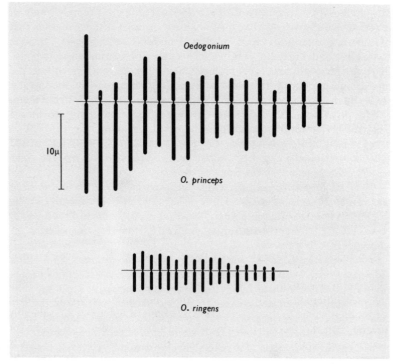

Fig. 3.1 Karyotypes of two different species of the algal genus *Oedogonium*, *O. princeps* (above) and *O. ringens* (below), shown diagrammatically, to illustrate differences in chromosome size. (Redrawn from L. R. Hoffman.[105])

groups. Certain genera of green algae, such as *Oedogonium*,[105] include some species with relatively large chromosomes and others with much smaller ones (Fig. 3.1). Among the fungi, most groups have very small chromosomes and low amounts of DNA in their nuclei, but certain of the cup fungi (Pezizales) among the Ascomycetes have much larger ones.[241] Nevertheless, the modal chromosome size and DNA content among algae and fungi are less than among bryophytes, and these, in turn, are less than the modal values for vascular plants.

On the other hand, both the mean and modal values for DNA content in different groups of vascular plants are rather similar, and the variation within classes, orders, and even families[11] may be far greater than the overall differences between classes. Some of this variation is due to multiplication of basic chromosome numbers, or polyploidy, and is discussed in later chapters. Nevertheless, great differences in chromosome size exist between genera of seed-bearing vascular plants. These can be seen in the Angiospermae by comparing two members of the pea family (Leguminosae), *Vicia faba* with *Lotus tenuis* (Fig. 1.2). The only group of plants which consistently has medium sized or large chromosomes is that of non-angiospermous seed plants: Conifers and Taxads, *Ginkgo*, Cycadales, and Gnetales. The various classes of marine algae, Rhodophyceae, Phaeophyceae, and marine Chlorophyceae, may constitute the only groups of plants with consistently small chromosomes, but they are not well enough known cytologically for generalizations to be made about them.

The distribution of these size differences is, however, by no means haphazard. Certain regularities must be noted, as follows.

(1) *Among spore bearing vascular plants, the heterosporous genera and families* (Selaginella, Isoetes, *Marsiliaceae, Salviniaceae*) *tend to have smaller chromosomes than the homosporous ones.*

(2) *Gymnosperms (Cycadales,* Ginkgo, *Coniferales) have the largest mean and modal chromosome sizes of any major groupings of plants.* Certain genera and families of angiosperms (*Paeonia, Lilium, Trillium, Tradescantia*, Krameriaceae, Loranthaceae) may, however, exceed gymnosperms in this respect.

(3) *Woody angiosperms mostly have small chromosomes with little difference in size between related species or genera.* A few exceptions to this rule exist, particularly the family Annonaceae of the order Ranales.

(4) *Among herbaceous angiosperms, great differences in chromosome size exist between different genera of the same family.* Groups showing these differences include both annual and perennial species. In some instances, as in *Oxalis*,[160,161] great differences in chromosome size exist between different species of the same genus. Moreover, there is no obvious relationship between the phylogenetic position of a family and the distribution of chromosome sizes within it. This is shown by Table 3.2.

Table 3.2 Contrasting sizes of somatic chromosomes in species of angiosperms belonging to the same family.

Family	Species	Gametic number	Mean chromosome length in microns
Ranunculaceae	*Isopyrum fumarioides*	$n = 7$	1.3
	Anemone hepatica	$n = 7$	7.0
Oxalidaceae	*Oxalis cuneata*	$n = 6$	1.5
	Oxalis dispar	$n = 6$	15.1
Leguminosae	*Lotus tenuis*	$n = 6$	1.8
	Vicia faba	$n = 6$	14.8
Droseraceae	*Drosera rotundifolia*	$n = 10$	0.8
	Drosophyllum lusitanicum	$n = 6$	14.0
Compositae, Tr.	*Agoseris heterophylla*	$n = 9$	2.4
Cichorieae	*Chaetadelpha Wheeleri*	$n = 9$	6.4
Liliaceae	*Tofieldia nuda*	$n = 15$	1.3
	Lilium pardalinum	$n = 12$	20.0
Gramineae	*Chloris barbata*	$n = 10$	1.5
	Secale cereale	$n = 7$	7.2

Hypotheses to explain differences in size

As explanations for these differences in chromosome size and nuclear DNA content, four géneral hypotheses can be suggested.

(1) A higher content of nuclear DNA indicates the presence of a larger number of active genes, and hence a greater diversity of gene function.

(2) Organisms having large amounts of nuclear DNA contain a high proportion of DNA with 'nonsense' sequences of nucleotides, which are devoid of any genetic function.

(3) Organisms with large amounts of DNA contain chromosomes which are replicated in their entirety, i.e., multistranded, and so are the genetic equivalents of polyploids, even though they appear to have diploid chromosome numbers.

(4) Organisms with large amounts of DNA contain gene loci which are replicated many times in series, tandem fashion. These replications might serve various functions. One of these could be to make possible the simultaneous transcription at a particular stage of development of large amounts of a particular kind of messenger RNA. This would greatly increase the rate of synthesis of the protein for which the messenger was coding. Another possibility, suggested by the present author,[217] would be that each member of a duplicated series could be activated in sequential order. The number of replicated units would, therefore, determine the length of time over which a particular protein was being synthesized.

1. *The hypothesis of added gene functions*

These hypotheses will now be considered in turn. The hypothesis of added gene functions may well explain much of the difference in DNA content between relatively little differentiated eucaryotes, such as *Chlorella*, *Euglena*, and *Chlamydomonas*, on the one hand and the earliest vascular plants on the other. In order to produce the great variety of cells and tissues which these latter plants contain, a relatively large number of genes must be required. Comparative studies of enzyme molecules indicate that a considerable proportion of this increase could be due to added gene functions. The evidence is based upon comparisons of the exact amino acid sequences of readily available animal proteins, such as the haemoglobin of the blood, and the antibody proteins of vertebrates. The best known example is the similarity between the two polypeptide chains, alpha and beta, of adult haemoglobin, the polypeptide chain of foetal haemoglobin, and that of myoglobin found in muscle tissues.[110] These molecules resemble each other so closely that they must have been derived from a common ancestral molecule, which underwent several tandem duplications during the course of evolution. Each duplication was then followed by differentiation in respect to portions of the amino acid sequences as well as the function of the molecule.

The data on amino acid sequences provide strong evidence in favour of an hypothesis conceived by geneticists many years before this information was available. The argument runs as follows. Since only a small proportion of the amino acid sequences which are possible will lead to a protein molecule having any kind of enzymatic function, the alteration of a functional molecule step by step as a result of mutations in the gene which codes for it could give rise to a molecule having a new function only via a succession of intermediate, non-functional states. If the gene being altered by mutation were the only one which could code for the enzyme concerned, then these intermediate steps would lack the necessary enzyme, and so would be cell lethal. Hence the new functional molecule could never appear. If, however, two or more genes exist which can code for the same kind of enzyme, then one or more of them can mutate more or less at random, while the other maintains the essential function, so that eventually the 'valley' between the two adaptive 'peaks' can be crossed. The evidence from homologous gene sequences in related proteins is now so extensive that this hypothesis is coming to be regarded as the principal way of explaining the evolutionary origin of new enzymatic functions.

The first hypothesis, therefore, that of added gene functions, serves well to explain at least some of the increase in chromosomal volume from bacteria to the earliest eucaryotes, and from them to the earliest vascular

plants. It is, however, completely incapable of explaining the wide variations in DNA content which exist within the various classes of plants.
These variations between different genera of the same family or even
between species of the same genus, as in legumes, grasses, *Oxalis* and
Oedogonium, are not correlated at all with structural complexity, and there
is no evidence to indicate a relationship with functional or enzymatic
complexity. It would be very hard to imagine why a broad bean (*Vicia
faba*) would require five times as many gene coded functions as a soy bean.
Similarly, cultivated rye has several times as much DNA in its nuclei as
panic grass (*Panicum* spp.). At a higher taxonomic level, chromosome
volume per nucleus, far from showing an increase as we proceed from
primitive spore-bearing plants (*Psilotum, Tmesipteris, Lycopodium*) through
more primitive seed plants (Cycads, *Ginkgo*, conifers) to the flowering
plants, is, on the average, somewhat less in angiosperms than in Psilotales
and lycopods. Clearly, other explanations than added gene functions are
needed.

2. *The hypothesis of 'nonsense' DNA*

The second hypothesis, that organisms with large chromosomes
possess large amounts of DNA with 'nonsense' sequences having no
adaptive value, is also inconsistent with the available data. If this were the
case, then we would expect that, over a long period of time, species which
had initially acquired a large amount of DNA would gradually lose it through
random deletions. Such species should exhibit both a great amount of
variability between individuals in DNA content, and perhaps even greater
differences between related species which had been isolated from each
other for long periods of time. This, however, is not the case. The genus
Trillium, for example, contains related species in temperate eastern North
America and eastern Asia, as well as in western North America. These
species are everywhere associated with the ancient, Arcto-Tertiary
deciduous forest, which has existed almost unchanged since early in the
Tertiary period.[209] On this basis, we can hardly escape the conclusion that
the genus *Trillium* is at least from forty to fifty million years old. Yet all
of the diploid species of *Trillium* agree with each other in having five pairs
of very large chromosomes. Both chromosome size and gross morphology
are remarkably constant throughout the genus. It is hard to see how this
constancy could have been maintained for such a long time unless it has
some kind of an adaptive significance. The same kind of argument can be
applied with equal or greater force to the genus *Pinus*, which is known from
fossil evidence to date back at least to the Cretaceous period, a hundred
million years ago. All species of pines are alike in having twelve pairs of
rather large chromosomes with similar morphology.

3. *Hypotheses involving gene duplication*

The third and fourth hypotheses both assume that organisms having large amounts of nuclear DNA possess some or all of their gene loci duplicated many times, and that these duplications are somehow adaptive. They differ only in their postulates as to how this duplication occurred. Consequently, the validity of either of them will be affected by evidence regarding the presence in higher organisms of duplicated gene loci, as well as indications as to how these duplications might adapt them to certain kinds of environments.

Several lines of evidence indicate that gene loci duplicated in tandem fashion exist in higher animals. Biochemical evidence comes from experiments in which the DNA was broken into fragments, which were then heated to separate their two strands or helices from each other. The rate of reassembly of these 'denatured' fragments was then followed during a carefully controlled cooling process.[23] The rationale of these experiments is that if the DNA contains many denatured fragments having similar nucleotide sequences, each single strand can unite with any of a number of matching partners, and reassembly will be relatively rapid. If, on the other hand, a gene locus is present only once, a denatured strand of this unduplicated DNA will be much less likely to find a matching partner, and the recovery of double-stranded DNA will be slow.

The outcome of these experiments indicates that in the nuclei of calf livers, about 40 per cent of the DNA consists of sequences which are repeated at least 10 000 times, and some sequences may be repeated a million times.

4. *The multiple strand hypothesis*

Evidence that the large chromosomes found in some genera of plants consist of chromonemata, including double helices of DNA, present in duplicate or quadruplicate, consists chiefly of direct observations of such chromosomes, and was reviewed in the last chapter. Particularly significant in this connection is the observation of subchromonemata in the large chromosomes of *Vicia faba*, and the failure to detect comparable subdivisions in the smaller chromosomes of *V. sativa*.[246]

If we accept as valid the evidence for multiple-strandedness in the largest chromosomes of higher plants, we should not dismiss the possibility that tandem duplications also exist. Even *Vicia sativa*, in which the chromosomes do not appear to be multistranded, has larger chromosomes than most species of the family Leguminosae. Furthermore, genera like *Crepis* include a wide range of chromosome sizes,[6] but these do not appear to be multiples of each other, as would be expected on the hypothesis that only duplications of entire chromonemata have taken place. For higher plants as a whole, hypotheses (3) and (4) presented above must be regarded as complementary to rather than competitive with each other.

5. *The tandem duplication hypothesis*

One advantage of the tandem duplication hypothesis as the most general explanation for differences in chromosome size between species of higher plants is that it can explain reversals of the trend toward increasing chromosome size. The existence of evolutionary trends from more generalized species having relatively large chromosomes to highly specialized species having smaller chromosomes was first pointed out by Delaunay[49] in the genus *Muscari* of the Liliaceae. In *Crepis*, Babcock[6] described a number of examples, which consist of parallel and independent trends from the perennial toward the annual life cycle, accompanied by reduction in the size of flowers and fruits, increasing specialization of the latter, and reduction in both size and number of chromosomes. Similar trends exist in other genera of Compositae, such as *Leontodon*, *Agoseris*, *Microseris*, *Eupatorium*, *Eriophyllum*, and *Haplopappus*. Examples in other families are *Knautia* (Dipsacaceae[120]) and *Phalaris* (Gramineae). These trends could be explained by assuming that evolution of annual growth cycles reduces the adaptive value of extra duplications of certain gene loci, so that they can be lost with impunity. In *Muscari*, of which the species are all bulb-forming perennials, the more specialized species having smaller chromosomes are characterized by reduced inflorescences and flowers, and in these respects resemble the reduced annuals of the Compositae.

VARIATIONS IN THE AMOUNT AND DISTRIBUTION OF HETEROCHROMATIZATION

The distribution of condensed or heterochromatic regions of chromosomes in metabolic nuclei was carefully studied on a comparative basis by Emil Heitz[101,102] and other cytologists in the 1930's and 1940's, but such comparative studies have been much less popular recently. The recognition that heterochromatization is closely associated with stage-specific gene repression and activation, as explained in the last chapter, should be a stimulus for a revival of interest in this field.

The earlier studies showed that heterochromatic regions may either be scattered irregularly over the chromosomes, or localized at various positions. The principal kinds of localization are: (1) concentration of heterochromatic regions near the centromere, forming prominent chromocentres or 'prochromosomes', which are often of the same number as the diploid chromosomal complement; (2) terminal knobs or satellites; (3) heterochromatization of entire chromosome arms or even entire chromosomes, often associated with sex differentiation.

Species with heterochromatic chromocentres

Species having heterochromatic chromocentres are scattered throughout the plant kingdom. In some instances they occur as relatively isolated examples in genera of which the other species have irregularly distributed heterochromatic regions. Typical examples are *Hordeum vulgare, Sorghum vulgare* and its close relatives, *Crepis fuliginosa*[231] and the closely related species *Plantago ovata* and *P. insularis*.[109,218] (Fig. 1.3). Examples of genera which consist largely or entirely of species having well marked chromocentres are *Scorzonera* (Compositae), *Oenothera* (Onagraceae), and *Collinsia* (Scrophulariaceae). Three families in which the majority of the genera include only species having well marked chromocentres are Onagraceae,[127] Solanaceae and Scrophulariaceae. The Ranunculaceae are sharply divided into two groups, one of which has large chromosomes with irregularly distributed heterochromatic regions, and the other (*Coptis, Isopyrum, Aquilegia, Thalictrum*) has small chromosomes with well marked chromocentres.[128]

Comparisons between the morphological characteristics and the systematic position of species having well marked chromocentres and related species which lack them indicate that at least in many instances the presence of these bodies represents a derived condition. *Hordeum vulgare* and *Plantago ovata* are both specialized annuals. Both the genera *Hordeum* and *Plantago* contain numerous other less specialized perennial species which have irregularly distributed heterochromatic regions. In the family Onagraceae, the genera *Lopezia* and *Fuchsia* which lack prominent chromocentres are regarded as less specialized and closer to the tropical woody ancestors of the family than the majority of the genera which have them.[127] *Scorzonera* is specialized compared to other related genera of Compositae, tribe Cichorieae, in respect to its linear leaves and tendency toward xerophytism.

The adaptive value of localized heterochromatic chromocentres may become apparent when more is known about the genes contained in them. A reasonable hypothesis about this adaptiveness has, however, been advanced on the basis of the expected genetic properties of these regions. Crossing over is known to occur only rarely in heterochromatic regions, so that all of the genes contained in the chromocentric region of any particular chromosome are much more closely linked than are genes situated at comparable distances from each other in other regions of the chromosome. Furthermore, as explained in the last chapter, the relaxation of the condensed regions of chromosomes tends to occur at the same developmental stage if these regions are closely adjacent to each other. Hence the genes contained in chromocentres would be expected to be all transmitted as a single Mendelian unit or 'supergene,' and would probably be activated more or less simultaneously at the same developmental stage. We would expect, therefore, to find well marked chromocentres or 'prochromosomes' evolving in response to the adaptive value of groups of genes capable of acting in coordination with each other at restricted stages of development.

Heterochromatic regions and sexual differentiation

The extensive heterochromatization which accompanies sexual differentiation is much more highly developed and better known in animals than in plants. In some groups of plants, however, this trend can be observed. The largest numbers of examples are in the Bryophytes, particularly the liverworts. The haploid thalli of these plants are in some species monoecious, with archegonia and antheridia borne on the same thallus, and in other species dioecious, including female thalli which bear archegonia and male thalli which bear antheridia.

Sex chromosomes of Bryophytes

The two prinicipal orders, Jungermanniales and Marchantiales, both contain monoecious as well as dioecious species. The monoecious condition exists in the genera *Calypogeia, Diplophyllum, Fossombronia* and *Lejeunea* in the Jungermanniales, as well as in all genera of the families Ricciaceae, Targioniaceae, and Rebouliaceae plus four genera of Marchantiaceae in the Marchantiales. All of these groups have karyotypes characterized by a minimal amount of heterochromatization. The smallest chromosome of the complement is partly or entirely heterochromatic in all of them, and in most of them one or more of the larger chromosomes possess heteromatic regions near the centromere or on their short arms (Fig. 3.2).

In both orders, the differentiation of the more specialized dioecious species is accompanied by increasing heterochromatization of certain chromosomal regions, but its course has been different. In the dioecious genera of Marchantiales (*Conocephalum, Marchantia*), only the smallest chromosome is affected. Where it can be detected, the difference between the sexes consists of a small dimorphic pair, the Y chromosome being smaller than the X (Fig. 3.3). In the Jungermanniales, on the other hand, the thalli of the dioecious species in nine different families are differentiated on the basis of the degree of heterochromatization of the largest chromosome. This is greater in the male than in the female (Fig. 3.2), although the overall size and form of the chromosomes are the same in both sexes. The genera *Pellia* and *Sphaerocarpos* each has a dimorphic X-Y pair, which has arisen through differentiation of this largest chromosome, probably in another direction.

In the Jungermanniales, the dioecious species contain more heterochromatic regions than the monoecious species in two or more of their

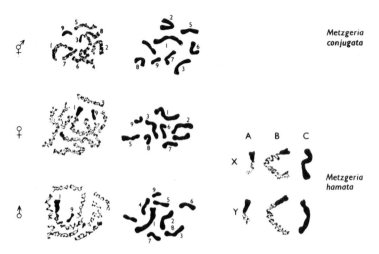

Fig. 3.2 Somatic chromosomes at prophase and metaphase in gametophytes of two species of the liverwort genus *Metzgeria*. Top row, *M. conjugata,* a bisexual or monoecious species. Lower two rows, female and male gametophytes. *M. hamata,* a unisexual or dioecious species. Note that the chromosomes of the female gametophyte are indistinguishable from those of *M. conjugata*, but that in the male gametophyte the largest or Y chromosome has a longer heterochromatic arm. (From Segawa.[193])

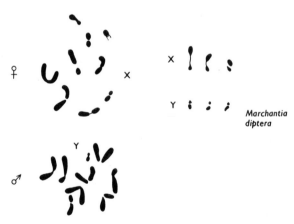

Fig. 3.3 Somatic metaphase chromosomes in the unisexual liverwort *Marchantia diptera*. In this species, the smallest chromosome pair has been converted into a heteromorphic X–Y pair. (From Tatuno.[227])

larger chromosomes, but differences between the sexes in this respect are confined to the largest chromosome of the complement (Fig. 3.2).

Because of the close relationships between their monoecious and dioecious species, as well as their favourable chromosomes, the liverworts are among the most promising organisms in either the plant or animal kingdoms for intensive cytological and genetical studies of the origin of sexual differentiation. For instance, the relatively primitive genus *Metzgeria* contains both monoecious and dioecious species. Furthermore, the karyotype of the monoecious *M. conjugata* agrees almost exactly in chromosome number, morphology, and degree of heterochromatization with that of the female gametophyte of the dioecious species *M. hamata*.[193] One might be inclined to postulate that the latter species has arisen from an ancestor similar to *M. conjugata* by a course of evolution which included increasing heterochromatization of the largest chromosome in the complement.

In one species of flowering plants, *Rumex thyrsiflorus*,[253] differential heterochromatization of Y chromosomes has been found. This species, like *R. hastatulus* as described in the next chapter, has two X chromosomes in females and in the males has one X and two Y's, that form a trivalent chain at meiosis. In nuclei of somatic cells, the two Y chromosomes are heterochromatic, and their DNA is replicated later in the cell cycle than that of either the autosomes or the X chromosomes.

Sex chromosomes in seed plants

In seed plants, the evolution of separate sexes, or dioecism, has occurred repeatedly. Examples such as *Ginkgo*, *Taxus*, the families Caricaceae, Salicaceae, Begoniaceae, Datiscaceae, and Dioscoreaceae, the genera *Cannabis*, *Humulus*, *Bryonia*, *Antennaria*, *Distichlis* and *Buchloe*, and certain species of such genera as *Thalictrum*, *Lychnis* (*Melandrium*), *Rumex* and *Mercurialis* are well known. An important fact is, however, that all of the families involved are relatively small, the dioecious genera are usually related to others that are monoecious or that have perfect flowers, while dioecious species belonging to genera that also contain monoecious or perfect flowered species usually are in the minority within their genera. This indicates that, under certain circumstances, separation of the sexes may have an immediate advantage for some species of seed plants, but that it rarely if ever increases their opportunities for later evolutionary diversification. This is in contrast to the wide-spread and general occurrence of separate sexes in animals. Moreover, sex chromosomes that are clearly recognizable on the basis of external morphology are the rule in animals but relatively uncommon in plants. These are discussed in the next chapter, since they contribute to the morphological differentiation of the karyotype.

The situation that was described above for liverworts might lead one to suspect that seed plants that lack morphologically distinguishable sex chromosomes nevertheless possess differential heterochromatization of one of their chromosome pairs. Unfortunately, this condition is very difficult to detect in species having chromosome numbers as high as those found in most dioecious seed plants. If present, it has not yet been recognized, except in *Rumex*, as mentioned above. By use of the techniques now available for spreading and staining chromosomes, differential hetero-chromatization might be detected by examining various species that have been previously studied only by older methods. More karyological research of this kind is needed.

POLYMORPHISM FOR ALLOCYCLY IN POPULATIONS

Allocycly was defined in the last chapter as the persistence of uncondensed chromosomal segments into stages, particularly late prophase and metaphase, at which the entire chromosome is normally condensed. In order to reveal it, low temperatures or some extreme environmental stimulus must be administered artificially. In plants with large chromosomes, variation with respect to this allocyclic response has proved to be a useful indicator of chromosomal polymorphism in populations. The principal research in this field has been carried out on species of *Trillium* by Japanese workers.[126] In *T. kamtschaticum*, a species which is common in northern Japan, they have found on the five chromosomes of its gametic set respectively 7, 19, 15, 16, and 18 different patterns or types of allocycly, differing with respect to the size and position of the uncondensed regions (Fig. 3.4). These differences probably reflect deficiencies and inversions of minute chromosomal segments. The number of variants which arise via such chromosomal changes is, however, probably much fewer than the number of types which can be recognized in populations, since new types can arise through crossing over.

An intensive study of 40 different populations of *T. kamtschaticum* distributed throughout northern Japan has revealed marked regularities with respect to both the amount of chromosomal variation in populations and the distribution of certain variants. The large populations which occur in the eastern part of the island of Hokkaido, which have been isolated from each other for a relatively short period of time, have the greatest amount of variability, while populations in northern Hokkaido and northern Honshu, which represent respectively the northern and the southern limit of distribution for the species in Japan, and which are relatively small and more completely isolated from each other, are much less variable. In respect to differences in frequency of particular chromosomal types between neighbouring populations, the small populations of the northern

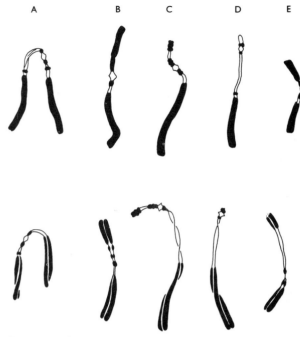

Fig. 3.4 Karyotypes of plants from two different populations of *Trillium kamtscha-ticum* of Japan, showing variation in allocyclic properties as revealed by cold treatment. (From Kurabayashi.[126])

region are relatively homogeneous, while those of the southern region are much more heterogeneous. The large populations of the eastern region are similar with respect to the frequency of certain of the most common types, but differ with respect to other types. On a regional basis, differences in the frequency of chromosomal types are more pronounced, and show a regular pattern. This made possible the division of the species into three regional groups, southern, northern, and eastern, each characterized by a distinctive array of chromosomal types (Fig. 3.5). This division coincides well with the distributional pattern of certain morphological characteristics. Comparable results with respect to both the amount of intra-populational variability and the regional differentiation of populations have been found in *Trillium ovatum* of western North America[76] (Figs. 3.6, 3.7).

The distributional pattern of chromosomal types in these species of *Trillium*, which is correlated with differences in respect to population size, regional and local climate, as well as geological history, is interpreted

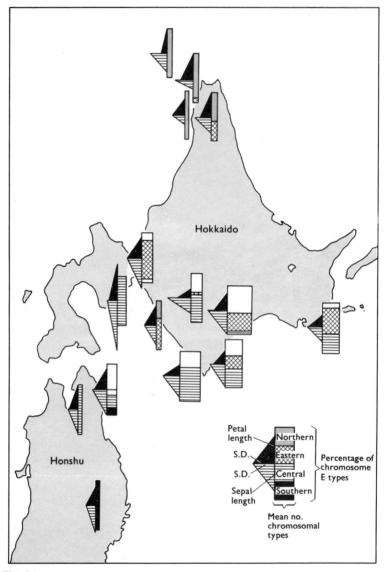

Fig. 3.5 Distribution of morphological and chromosomal variants of *Trillium kamtschaticum* in northern Japan. The symbols represent mean values and amounts of variation with respect to two morphological characteristics (left) and allocyclic variants of somatic chromosomes (right). The width of the base of each triangle along the median base line represents the amount of standard deviation in the population for petal length (top triangle) and sepal length (bottom triangle). The height of each triangle represents the mean values in each population for petal length and sepal length. The width of the rectangle on the right side of each symbol is proportional to the number of chromosomal types found in the population, while the different kinds of shadings represent the percentage frequency in the population of each of the most common chromosomal types. (From Kurabayashi.[126])

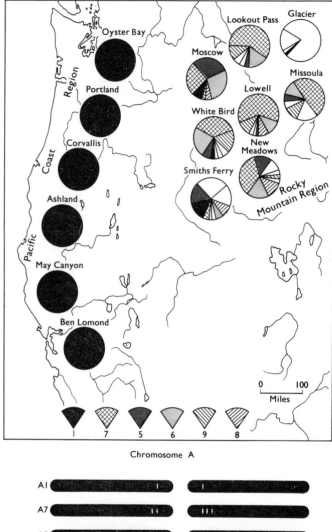

Chromosome A

Fig. 3.6 Map showing the distribution in populations of *Trillium ovatum* in the western United States of six different allocyclic variants of the large metacentric Chromosome A. Note the contrast between the uniformity of Pacific Coast populations and the heterogeneity as well as geographic variability of populations from the Rocky Mountains. (From Fukuda.[76])

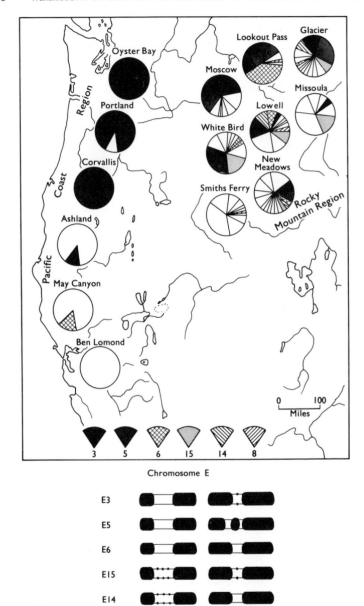

Fig. 3.7 Map showing the distribution in the same populations as those shown in Fig. 3.6 of allocyclic variants of the E chromosome of *Trillium ovatum*. With respect to this chromosome, Pacific Coast populations show some heterogeneity, and a definite pattern of geographic variability, although in both respects the populations are more constant than those of the Rocky Mountains. (From Fukuda.[76])

as a result of complex interactions between mutation, genetic recombination through crossing over, natural selection, and genetic drift. This example shows that when they are available, chromosomal differences are just as valuable as gross morphological characteristics to serve as markers for studying the dynamics and evolution of natural populations. Moreover, chromosomal differences may reflect differential gene action more closely than external morphology, so that when they are better understood they may provide a valuable link between the action of selection on the phenotype, and the response of the population in terms of genotype frequency. They deserve more attention for this purpose, particularly if differences can be found in a species which can be relatively easily handled genetically.

DISTRIBUTION AND SIGNIFICANCE OF ACCESSORY CHROMOSOMES

A final feature of many plant species which must be considered in connection with heterochromatization and gene action is the presence and behaviour of accessory chromosomes. These chromosomes, which have also been called B chromosomes and supernumerary chromosomes, are now known to exist in more than 150 species of flowering plants, distributed over a wide range of families, as well as in many species of animals.[171] They differ from the basic chromosomal complement chiefly with respect to inconstancy in number, smaller size, and greater degree of heterochromatization (Fig. 3.8).

The number of accessory chromosomes may differ from one tissue to another of the same organism, and from one individual to another of the same population. Regional differences with respect to the mean number of accessories in a population occur in many species.

Variation within an individual results from the systematic elimination of accessories from certain organs, and more rarely their increased rate of replication in particular tissues. Accessories are found in pollen mother cells and other reproductive tissues in *Poa alpina, P. timoleontis, P. trivialis, Sorghum purpureo-sericeum, S. nitidum, Panicum coloratum, Xanthisma texanum, Haplopappus gracilis,* and *H. spinulosus* ssp. *cotula.*[13] In most of these species, the distribution of accessories among the various tissues of the shoot has not been studied, but in *Poa alpina* they are known to be absent also from the leaves, while in *Xanthisma texanum* they are present in various tissues of the shoot. On the other hand, the number of accessory chromosomes is remarkably constant in all tissues of the plant in many species, such as those of *Agrostis, Dactylis, Festuca, Secale,* and other genera of grasses.

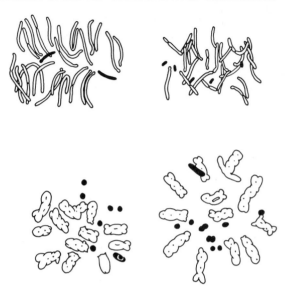

Fig. 3.8 Somatic (top row) and meiotic (bottom row) metaphase plates in *Lilium medeoloides,* showing various combinations of accessory chromosomes. (From Samejima.[189])

In gross morphology, most accessory chromosomes are acrocentric or telocentric. Those which are metacentric are usually if not always iso-chromosomes, in which the two arms are duplications of the same arm, and so are homologous with each other.

At meiosis, homologous pairing between accessory and basic chromosomes does not usually occur. A few exceptions such as in *Clarkia*, have been recorded. On the other hand, when accessories are present in even numbers they often, though not always, pair with each other. This pairing is most easily seen at pachytene. Generally, it is not accompanied by the formation of chiasmata, and the chromosomes may become separated from each other at later prophase and first metaphase.

Morphological and physiological effects

No example is known of accessory chromosomes which have recognizable effects upon the external morphology of the plant. This is evident from the fact that in populations or artificial progenies containing variable numbers of accessories, individuals having higher numbers are not generally distinguishable from those which have fewer.

The principal physiological effects are on the overall vigour of the plant, as well as pollen fertility. Vegetative vigour is usually reduced only by high numbers of accessories, such as ten or more in *Zea mays*, but in some species, such as cultivated rye (*Secale cereale*), a noticeable decrease in vigour occurs in plants which have only one or two accessories. This effect is very pronounced in artificial autopolyploids, and it is a general rule that polyploids are much less likely to contain accessory chromosomes than their diploid relatives. Pollen fertility is more sensitive to the presence of accessories than other physiological characteristics which have been studied. In *Sorghum purpureo-sericeum* any increase in the number of accessories produces a reduction in pollen fertility. Other species are less sensitive, such as *Centaurea scabiosa*, in which fertility is not reduced until more than five or six accessories are present.

Favourable effects of accessories are sometimes found. In *Festuca pratensis* vigour may sometimes be increased by the presence of one or two accessories. A rather unusual situation exists in diploid species of *Achillea*, *A. asplenifolia* and *A. setacea*.[61,62] Plants of these species having two accessories are vigorous, and may be more fertile than plants which lack them altogether. Moreover, in the cross *A. asplenifolia* × *setacea*, though not in its reciprocal, F_1 plants could be obtained only when the female parent contained accessories. In this instance, the presence of accessories in the stylar tissue of *A. asplenifolia* apparently promotes the growth of pollen tubes of *A. setacea*. On the other hand, plants of species or hybrids of this group which contain uneven or high numbers of accessories are relatively infertile. Moreover, such plants show considerable instability in respect to the number of accessories found in various tissues.

Transmission and elimination of accessories

The mean number of accessories present in a population or a species may be altered by their increase or decrease during the development of the germ line, particularly the archesporium, and/or during gametophyte development. In addition, their number may be kept in check by selective elimination of individuals having high numbers of accessories.

Increase during the development of the germ line takes place in *Crepis*[74] and *Achillea*.[62] The mechanism for this increase has not been demonstrated. In some other genera, such as *Lilium, Trillium, Tradescantia*, and *Plantago*, univalent accessories appear in unexpectedly high frequencies in egg cells due to their preferential distribution to the functional megaspore at meiosis in the ovules. Much more often, however, their increase results from preferential distribution during the mitoses in the male gametophytes, or developing pollen grains. In the largest number of known examples (*Festuca, Briza, Holcus, Alopecurus, Phleum, Anthoxanthum, Collinsia*), the division involved is the first one, which gives rise to the generative and the vegetative nucleus. At this division the centromeres of univalent accessories divide as expected at pro-metaphase, but the proximal portions of the chromosome arms remain close to each other. Consequently, the two daughter chromosomes do not separate from each other, but pass together to that pole of the spindle which will form the generative nucleus and eventually the two sperm nuclei. Their separation during the metabolic stage, and their subsequent independent and normal behaviour during the second gametophytic division, cause the sperm nuclei to contain two accessories when only one was present in the somatic cells.

A different kind of preferential distribution occurs in maize, at the second gametophytic division. In this plant, the two sperm nuclei differ in the number of accessories which they contain, and the one with the highest number apparently fertilizes preferentially the egg cell. A still more unusual situation exists in *Sorghum purpureo-sericeum*, in which the first pollen grain mitosis is normal, but the vegetative nucleus undergoes extra mitoses, characterized by preferential distribution of accessories to the resulting generative nuclei.

Obviously, the continuation of this process unchecked over many generations would give rise to individuals having very large numbers of accessories. The check which keeps their number down is the relative weakness and poor competitive ability of individuals possessing high numbers of accessories.

The evolutionary origin of accessory chomosomes is still largely a matter of hypothesis and conjecture. The fact that, with few exceptions, they show no signs of homology with chromosomes of the basic set, and are completely different in size and morphology, suggests that in most species which have them, they are not recently derived from the basic chromosomes.

The ecological significance of accessories

In some instances, as in *Festuca pratensis*,[20] *Centaurea scabiosa* and *Lilium medeoloides*, populations containing accessories exist in a restricted portion of the geographic range of the species which is associated with specific factors of environment or population structure. In Sweden, populations of *Festuca pratensis* having accessories occur chiefly in regions having soils with a high clay content, a condition which is favourable for the species as a whole. Nevertheless, in other parts of the range of the species where the soil is rich in clay, the populations of *F. pratensis* completely lack supernumeraries. On the basis of the equilibrium hypothesis, this difference could be explained by assuming that populations containing accessories either evolved in or were introduced into Sweden, and survived in habitats sufficiently favourable for the species so that the genetic tendency for accumulation of accessories was not overbalanced by selective elimination of plants bearing them. In Britain, one could postulate either that accessories were never introduced into British populations, or that the mechanisms for the perpetuation in these populations were too weak to overcome the selective elimination.

In the case of *Lilium medeoloides* high numbers of accessories are found in populations which reproduce frequently by sexual means, so that preferential genetic segregation would have a high probability of being effective, while populations lacking accessories are much more stable, and reproduce very little by sexual means. In these latter populations, the combination of somatic elimination of accessories during asexual reproduction by bulblets, and selective elimination of plants bearing accessories is apparently responsible for their absence.[189]

With respect to the major processes of speciation and evolution, accessory chromosomes must certainly be regarded as of little fundamental significance. Nevertheless, their presence raises many interesting questions about chromosome mechanics, as well as the functional significance of certain genes. What forces in the cell cause supernumeraries to pass preferentially toward one pole of the spindle during certain mitotic divisions and not others? How can the property vary from one kind of accessory chromosome to another? What sorts of primary products do the genes located in accessory chromosomes produce, which have no visible effect on morphology, but appear often to have physiological effects? Can accessories ever contribute toward an increase in the basic chromosome number of a species, through translocation onto them of essential material from another chromosome? These and other questions are waiting to be answered by cytologists who have the ingenuity and persistence to design and carry out the right experiments for these purposes.

4

Chromosomal Changes, Genetic Recombination, and Speciation

In the first chapter of this book, the fact was emphasized that adaptive differences between individuals and populations are usually based upon interaction between many different genes. Because of this fact, an adaptive advantage can often be conferred upon mechanisms which transmit as units, clusters of interacting genes, causing each gene cluster to segregate and recombine as if it were a single Mendelian gene. Chromosomal changes which produce and alter the degree of genetic linkage between the genes which make up such adaptive clusters are, therefore, of great evolutionary significance.

CHROMOSOMAL REARRANGEMENTS AND THEIR EFFECTS ON LINKAGE

Two kinds of structural rearrangements of chromosomes have pronounced effects on genetic linkage, *inversions* and *translocations*. Inversions occur when a chromosome breaks in two places, and the segment between the breaks becomes turned around, so that the order of its genes is reversed with respect to that on the unbroken chromosome (Fig. 4.1). Spontaneous inversions probably occur through accidental breakage at the prophase of either mitosis or meiosis. Inversions have long been recognized by geneticists as suppressors of crossing over, and hence as

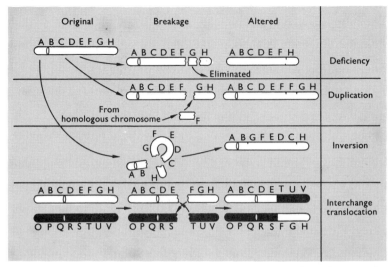

Fig. 4.1 Diagrams to show how chromosome breakage and reunion can give rise to the four principal structural changes which chromosomes undergo: deficiency, duplication, inversion and translocation. (From Stebbins.[216])

promoting genetic linkage. This is because crossing over depends upon precise gene-by-gene pairing at meiotic prophase. If this happens in an individual heterozygous for an inversion, the inverted segment must pair with its normal counterpart in such a way that a loop is formed at the meiotic prophase (Fig. 4.2). Crossing over can occur within this loop, but, when it does, the resulting crossover chromatids are abnormal and inviable. One of them is a fragment which lacks a centromere, and so cannot participate in mitotic division. The other is a dicentric chromatid, which at first anaphase of meiosis becomes stretched between the poles of the spindle. It then breaks, and the two broken, centric fragment chromatids are deficient for so many genes that cells which receive them are inviable.

The diagram of Figure 4.2 illustrates the effects of heterozygosity for **paracentric inversions**, which do not include the centromere. In the case of **pericentric inversions**, which include the centromere, crossing over within the inverted segment gives rise to a different behaviour of the meiotic bivalent, illustrated in Fig. 4.3. Its end results are, however, the same; inviability of the crossover chromatids.

A second effect of inversions is to bring together into the same chromosomal region genes which previously were widely separated from each other, thus increasing the amount of linkage between them. If, for instance, two genes occur on the same chromosome, but on opposite sides of the centromere, a pericentric inversion may bring them together as components of an adaptive gene cluster.

Distribution of inversions in populations

The significance of inversions in populations of insects, particularly *Drosophila*, has long been recognized.[55] In higher plants, they also occur, but in the absence of giant chromosomes in which precise gene-by-gene pairing can be easily observed, their nature and distribution cannot be

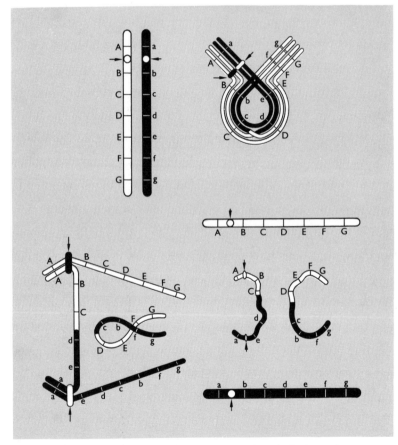

Fig. 4.2 Diagram to show how crossing over at meiosis in a chromosome heterozygous for a paracentric inversion can lead to the elimination of chromosome strands containing the recombined genes. (From Stebbins.[216])

accurately recorded. Nevertheless, in most species of higher plants bridge-fragment configurations similar to the one illustrated in Figure 4.2 are rare or lacking at meiotic anaphase, being frequent only in inter-specific hybrids. This fact would suggest either that polymorphism for inversions is much less common in populations of plant species than it

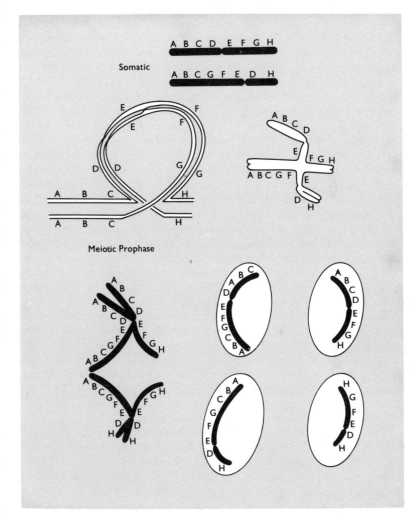

Fig. 4.3 Diagram to show how crossing over at meiosis in a bivalent heterozygous for a pericentric inversion can lead to two abnormal chromatids, one of which contains a duplication and the other a deficiency.

is in *Drosophila*, or that inversions, if present, are so small that crossing over rarely occurs within them.

The nature and occurrence of translocation heterozygotes

Translocations occur when two non-homologous chromosomes break simultaneously and exchange segments (Fig. 4.4). If a plant becomes homozygous for such a rearrangement, some of its genes have been transferred to a completely different chromosome, so that their linkage relationships are completely altered. This is represented in Figure 4.4 by genes E, F, S, and T. The close linkage between E and F, which existed before the translocation occurred, is replaced in genotypes homozygous for the translocation by an equally close linkage between E and T.

In addition, translocations which are maintained permanently in the heterozygous state may keep together combinations of genes located on non-homologous chromosomes and cause them to be inherited as if they were genetically linked. This condition is the inevitable outcome of the pairing of homologous chromosomal segments in translocation heterozygotes. As is shown in Figure 4.4, the four chromosomes involved in a single translocation pair to form a cross-shaped configuration at the pachytene stage of meiotic prophase, and this cross opens out to form a ring at late prophase and first metaphase. Of the three possible kinds of arrangements of this ring on the spindle at metaphase, only one will cause the chromosomes to segregate in such a way as to give rise to gametes with a balanced complement of genes, lacking duplications or deficiencies. This arrangement is such that each gamete receives one of the two combinations of centromeres which were present in the parents of the heterozygote. Gametes containing new recombinations of centromeres will inevitably be deficient for certain chromosomal segments and duplicated for others. Consequently, those genes which lie so close to a centromere that crossing over between them and this body is rare will not only be linked to each other, but will also be linked to genes lying close to the centromere of the other non-homologous chromosome which is involved in the translocation.

This device for promoting interchromosomal linkage can be successful only if the chromosomes have certain definite structural features. First, their centromeres must be in a median or submedian position. Heterozygotes for translocations involving acrocentric or telocentric chromosomes form complex configurations at meiotic metaphase which are not ring shaped, and in which adjacent chromosomes often pass to the same pole. Since combinations of two adjacent chromosomes always include duplications and deficiencies for chromosomal segments, a high proportion of gametic inviability or sterility results from such segregations.

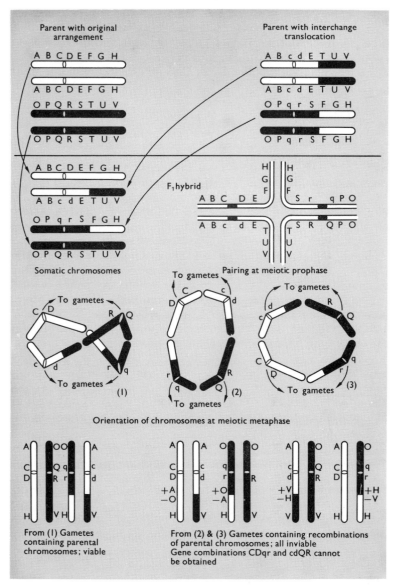

Fig. 4.4 Diagrams showing how homologous pairing in an organism hetero-zygous for a large chromosomal interchange can lead to a ring of four chromo-somes at meiosis, and how different orientations of the chromosomes in the ring at meiotic metaphase can lead to either viable or inviable combinations of chromo-somes in the gametes. (From Stebbins.[216])

In addition, chiasma formation must be confined to the chromosome ends, so that the ring configurations are flexible and can easily assume the zigzag shape which results in the distribution of alternate chromosomes to the same pole. This distribution of chiasmata may be promoted by the existence near the centromere of large heterochromatic regions, in which chiasmata are not formed.

These three conditions are met to the fullest degree in various genera of the Onagraceae, such as *Oenothera*, *Clarkia*, and *Gaura*. It is not surprising, therefore, that structural heterozygosity for translocations is more highly developed in these genera than anywhere else in the plant kingdom.

Interchange heterozygosity and ring formation in *Oenothera*

The anomalous chromosome behaviour in the subgenus *Euoenothera* of *Oenothera*, and its relation to the complex interrelationships between genotypes as well as the evolutionary success of the group, were the subject of much study during the 1920's and 1930's. The group is of classic significance because genetic mutations were first described in it by De Vries. In more recent years, the evolutionary processes which have given rise to its present cytogenetic structure have been largely revealed by the research of R. E. Cleland and his associates.[35,36,37]

Degrees of structural heterozygosity

All members of the group are alike in possessing a somatic complement of 14 chromosomes with median centromeres. They are all self compatible and capable of self fertilization, although self incompatibility exists in related species of *Oenothera*. In respect to the degree of structural hetero-zygosity, three modal conditions can be recognized. The first consists of plants with large flowers, which are often outcrossed, and which at meiosis form either pairs or small rings, due to heterozygosity for one or two interchanges. Populations of this kind are the only ones found in California and adjacent states, and are classified as the species *O. Hookeri*. Other populations which have large flowers and form pairs at meiosis, classified as *O. grandiflora*, occur in the Gulf coast of Alabama; and still others, under the name *O. argillicola*, on the shale barrens of the Appalachian mountain system. Some populations of these species are polymorphic for translocations, as are many other species in other subgenera of *Oenothera*, as well as the related genus *Clarkia*. Hence adaptive polymorphism for translocations already existed sporadically in the ancestors of the ring-forming *Oenotheras*.

The second condition, which is found in western North America from Colorado and Utah south into Mexico, is that of heterozygosity for numerous interchanges, so that all of the individuals of a population form chromosome rings of various sizes at meiosis. These rings, however, are not permanent, since progenies of these plants always contain some individuals with seven pairs. Like those of the first group, plants of this kind are large-flowered and open pollinated. Although experimental evidence for it is not available, the most probable explanation for the persistence of translocation heterozygosity in these plants is the adaptive superiority of the heterozygotes.

The third condition, which is the only one found in most of northern, central, and eastern North America, is permanent structural heterozygosity for translocations involving all of the chromosomes, so that rings of 14 are regularly formed at meiosis (Fig. 4.5). These extreme structural

Fig. 4.5 Prophase (above) and metaphase (below) of meiosis in a complex heterozygote of *Oenothera*, showing the ring of fourteen chromosomes. (From Cleland.[35])

heterozygotes are usually self pollinated and so are highly inbred. Not unexpectedly, seeds grown from any single individual produce offspring which exactly resemble their parent. Phenotypically, they appear like the individuals of a pure line of a typical self fertilized plant. Genotypically, however, they are entirely different. This is evident cytologically by the fact that all progeny form rings of 14, and so must be structural heterozygotes which have arisen through a union between two gametes having chromosomes with different structural arrangements. It becomes evident genetically when reciprocal crosses are made, either between two races of this type or between such a race and a structurally homozygous seven-paired *Oenothera*. In either case the reciprocal hybrids in most instances are entirely different from each other. If, for instance, a ring-forming race from the northeastern United States is crossed with a pair-forming race from California, the progeny of reciprocal crosses differ widely from each other, and resemble their maternal parent. Progeny of crosses between two ring-forming races having different morphology are also phenotypically maternal in the F_1 generation. The F_2 progeny of crosses between a ring former and a pair former are, however, entirely different from the F_2 of a cross between two ring formers. In the first case, they segregate for both morphological characteristics and form rings of various sizes, while progeny of crosses between ring formers usually breed true in later generations for both a particular combination of phenotypic characteristics and the consistent formation of rings of 14 chromosomes at meiosis.

Balanced lethals and gametic complexes

The cytological explanation for these facts lies both in the consistent orientation of the rings so that alternate chromosomes always go to the same pole, and in the presence of balanced lethal combinations of genes or chromosome segments which cause the death of zygotes which result from the union of similar gametes. A few races exist, including the classic *O. Lamarckiana*, in which only these two mechanisms are operating. Since they produce 50 per cent of abortive ovules, their reproductive potential is greatly impaired. The evolution of a more efficient reproductive mechanism has included the origin and establishment of two further complications. In most races of *Oenothera* the two possible kinds of gametes differ consistently in their effects on development. One set of chromosomes, known as alpha, is found in all of the functional egg cells. The opposite complex, designated beta, is eliminated from the developing megaspore tetrads, since haploid megaspore nuclei containing it consistently fail to develop. On the other hand, the beta complex is able to direct the development of the male gametophyte. It is present in all of the functional pollen grains, since microspores containing the alpha complex do not function. These abnormalities of development produce plants with 50 per cent of pollen abortion but a full seed set. Since the alpha complex of the egg always unites with the beta complex carried to it by the pollen, the plants breed true for the structurally heterozygous condition. Figure 4.6 is a set of diagrams illustrating these two kinds of balanced lethal mechanisms.

The variation pattern produced by such a genetic system is superficially similar to that found in ordinary self fertilizing species. Both contain a large number of closely related biotypes, represented in nature by thousands of similar individuals. Under ordinary conditions genetic segregation is much reduced. Structural heterozygotes differ greatly, however, from self fertilizers in their evolutionary responses to changes in the environment. When such changes bring together different genetically and chromosomally homozygous biotypes of a normal self fertilizing species, the result of occasional crossing between them is a short burst of evolution, in which complex segregation of genes in the F_2 and later

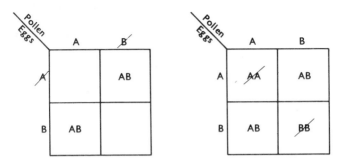

Fig. 4.6 Diagrams to illustrate the mode of action of two kinds of balanced lethal mechanisms in *Oenothera*. Left, gametophytic lethals, one of which (**A**) is lethal to female gametophytes and the other (**B**) to male gametophytes. Only zygotes of the constitution AB can be produced. Right, zygotic lethals, which cause the death of AA or BB zygotes, and bring about 50 per cent seed sterility. (From Cleland.[35])

generations presents to the environment a large number of new genotypes, which are sifted out by the action of natural selection. On the other hand, crossing between two biotypes of ring-forming *Oenotheras* produces, even in the F_2 and later generations, only two new biotypes, which often are much like the parental ones, but may be considerably different.

New variation by crossing over

The ring forming complex heterozygotes of *Oenothera* have, however, a capacity for giving rise to strikingly different variants in the absence of hybridization: a capacity which is lacking in ordinary self fertilizing biotypes. It is based upon the presence of proximal chromosomal segments, near the centromere, which are kept in a permanently heterozygous condition, and which usually do not undergo crossing over at meiosis. The union of the chromosomes at meiosis to form the ring is accomplished by chiasma formation and crossing over in their distal segments. Hence genes located in distal segments behave in the same way as do most of the genes found in a self fertilizing species. They may segregate independently of each other, and since all of their alleles, whether dominant or recessive, are eventually exposed to the action of natural selection, distal segments tend to become homozygous for the most favourable gene combination.

Proximal segments, on the other hand, are usually retained in the heterozygous condition by the absence of crossing over and the mechanisms which perpetuate the union of unlike gametes in zygote formation. This makes possible the accumulation of recessive mutations in them. Moreover, crossing over can occasionally take place in proximal or differential chromosome segments, and when it does, it has profound consequences, both cytologically and genetically. By various complex methods, such crossing over can give rise to completely new combinations of chromosome ends, which are associated with equally different morphological characteristics. The complex heterozygotes of *Oenothera* possess, therefore, a hidden pool of genetic variability, which can be released from time to time.

Evolutionary history of Euoenothera

The unusual cytogenetic mechanism for storing and releasing genetic variability which has been evolved by the ring-forming species of *Oenothera* has made possible a course of evolution characterized by repeated hybridizations and spread of hybrid derivatives.[35] At present, the ring formers can be divided into five groups (Fig. 4.7). One of these, known as *strigosa*,

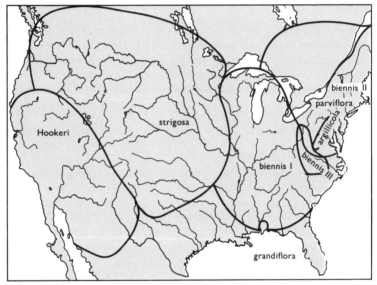

Fig. 4.7 Geographic distribution of the major cytological and morphological groupings of *Oenothera*, sect. *Euoenothera* in North America. (From Cleland.[37])

consists of biotypes adapted to the dry conditions of the Rocky Mountains and Great Plains. In them, both the egg (alpha) and pollen (beta) complexes carry genes for small flowers, narrow leaves, and hairiness. Three others, designated *biennis* I, II, and III, all possess chromosomal complexes which carry genes for broad, crinkly leaves and sparse pubescence and are adapted to the more mesic conditions which prevail in eastern North America. In *biennis* I, which occurs in the eastern part of the Great Plains, this latter complex is alpha, and is carried though the egg, while the pollen or beta complex carries *strigosa*-like characteristics. The races grouped under *biennis* II, and found in the north-eastern United States and adjacent Canada, have an alpha complex bearing *strigosa*-like characteristics and a beta complex bearing *biennis*-like characteristics. The more restricted group designated *biennis* III, and confined to the Appalachian mountain system, has complexes bearing *biennis*-like characteristics in both pollen and egg, and may have arisen through crossing between *biennis* I and *biennis* II. Finally the group designated *parviflora* has a beta complex carrying *biennis*-like characteristics combined with an alpha complex which bears characteristics similar to those found in the pair-forming race, *O. argillicola*, mentioned above.

The origin of these groups is attributed to a succession of migrations from an original centre in Mexico, followed by hybridization between new immigrants and older established populations. The first of these migrations was a race resembling the modern pair forming *O. argillicola*. It was followed by a later migration consisting of plants similar to the modern *O. grandiflora* of the south-eastern states. The chromosomal differentiation of these kinds of populations had already progressed far enough so that hybrids between them formed rings of 14, and had the beginnings of the balanced lethal system. These hybrids became the modern *parvifloras*. Two later migrations consisted of narrow-leaved, relatively xerophytic races similar to the modern *strigosas*. Hybridization between these races and the *grandiflora*-like races already present gave rise to the modern ring-forming populations, grouped under *biennis* I and II. The *strigosas* arose from hybrids between the last two groups of immigrants.

Oenothera-like systems in other plant groups

Although predominance of complex heterozygotes which form complete rings at meiosis is a rare situation in higher plants, it is not confined to the subgenus *Euoenothera*. Although less completely studied, the complex heterozygotes found in other subgenera of *Oenothera*: *Hartmannia*, *Lavauxia*, and *Raimannia*, are just as well developed as are those in *Euoenothera*. A similar situation has been uncovered by recent intensive studies of the genus *Gaura*, which, though distantly related to *Oenothera*, belongs also to the Onagraceae and has similar chromosomes. In *Gaura*, also, the evolutionary history appears to be one of successive migrations from a Mexican centre followed by hybridizations, and the trend from cross fertilization to self fertilization has also taken place.

Examples in other families, such as *Rhoeo discolor, Paeonia Brownii* and *P. californica,* and *Hypericum punctatum,* have long been known[209] but have not yet been subjected to intensive cytogenetic investigations, largely because they are technically more difficult to handle than are the members of the Onagraceae.

Origin of complex-heterozygote systems

Sufficient information is now available for constructing a reasonable hypothesis to explain how complex-heterozygote systems arise. The essential features of this hypothesis have been recognized for some time.[39] They may be summarized as follows. The ancestral populations were outcrossing pair formers, with normal sexual reproduction which ensured segregation and recombination of genes. They were probably heterozygous and polymorphic at many gene loci, as are most outcrossing populations. Moreover, many of them were also heterozygous and polymorphic for translocations, as are a large proportion of modern species of Onagraceae.

The cytological characteristics which favoured the accumulation of translocations were the presence of median centromeres, heterochromatic proximal chromosomal segments in which chiasma formation and crossing over were rare or absent, and the restriction of chiasma formation and free gene recombination to distal segments of chromosomes.

Given these favourable pre-adaptations, the impetus for the evolution toward permanent structural heterozygosity was probably provided by drastic oscillations of the climate combined with the elevation of mountain ranges, both of which were taking place in western North America during the Pleistocene epoch. These environmental changes made possible the alternate isolation of populations, accompanied in some of them by selection for adaptation to extreme environments, and their subsequent reunion, followed by hybridization. Some of this hybridization was probably between populations which, during a previous period of isolation, had acquired different translocations. In this way, vigorous hybrids heterozygous for a number of different chromosomal segments could have arisen. These could have given rise directly to populations characterized by many rings of intermediate size, such as are found in the *Euoenotheras* of the south-western United States and Mexico.

The step from such populations, heterozygous for many translocations but still capable of producing structural homozygotes by segregation, to permanent structural heterozygotes forming exclusively large rings, would be promoted by still another cycle of isolation followed by reunion and hybridization. During this cycle, mutation and gene recombination would have to act to produce the distinctive characteristics of the permanent structural heterozygotes. Mutations to self compatibility, which are well known in *Oenothera*, would eliminate any residual self incompatibility systems which might be present. Some of the self incompatibility alleles, on the other hand, could become incorporated into the chromosome sets or complexes which are transmitted only through the female gametophyte and egg and, by inhibiting the possibility for fertilization possessed by the pollen grains which carried them, could thus contribute to the balanced lethal system.[224] Other elements of the balanced lethal system would arise through the deficiencies and duplications for chromosomal segments which are the inevitable result of successive translocations. Based upon the known characteristics of monosomic plants, which are deficient for whole chromosomes, one would expect that some segmental deficiencies would be lethal in microspores but viable in megaspores and eggs, while others would be viable in microspores but lethal in female gametophytes. If many crossings between populations heterozygous for numerous translocations were taking place, some of them would be likely to have these complementary deficiencies distributed on the chromosomes in such a way that the regular behaviour of the permanent structural heterozygotes would be approximated. Once a moderately efficient mechanism of this sort had arisen in a vigorous, well adapted, and aggressive hybrid, natural selection for particular additional changes, brought about either by mutations, crossing over, or further translocation, could perfect it into the efficient system of permanent structural heterozygosity which we see today.

In the case of *Euoenothera*, the oscillations of climate which would have promoted this kind of evolution occurred repeatedly during the advance and retreat of glaciers during the Pleistocene, while a final union of previously isolated populations was brought about by the activity of man, since most biotypes of *Oenothera* are aggressive roadside weeds.

THE EVOLUTION OF KARYOTYPES

As explained in Chapter 1, the karyotype is the morphological aspect of the chromosome complement as seen at mitotic metaphase. Usually, somatic karyotypes are studied and compared, most often from root tip mitoses, but in certain favourable groups, particularly in the order Liliales and their relatives, gametophytic karyotypes from the first microspore division can be used. In the Bryophytes, gametophytic karyotypes are normally used.

The principal morphological criteria and their significance

Five different characteristics of karyotypes are usually observed and compared: (1) differences in absolute size of the chromosomes; (2)

differences in the position of the centromere; (3) differences in relative chromosome size; (4) differences in basic number; (5) differences in the number and position of satellites. A sixth characteristic, differences in degree and distribution of heterochromatic regions, can be studied if suitable mitotic prophases are available.

As explained in the last chapter, differences in absolute chromosome size between related species or genera probably reflect different amounts of gene duplication, either in tandem fashion or through polytene multiplication of chromonemata. These differences will not be discussed further.

The next three differences all reflect structural changes of the chromosomes. The position of the centromere can be altered by either pericentric inversions or unequal translocations. Differences in relative chromosome size can be brought about only by segmental interchange involving translocations of unequal size.

Aneuploid alterations of the basic chromosome number (see p. 17) are usually the outcome of successive unequal translocations, which progressively increase the differences in relative size between chromosomes. If the smallest chromosome of the complement also has heterochromatic regions near the centromere, in which the genes are either completely inactivated or act only during a restricted stage of development, then a final unequal translocation can remove all of the essential genetic material from this chromosome, so that its centromere can be lost without damage to the plant. This brings about an aneuploid reduction in the chromosome number. This hypothesis, known as the **dislocation hypothesis,**[209] is now supported by cytogenetic information in several genera.

Mechanisms for aneuploid increase of the chromosome number are of necessity more complex. Two such mechanisms probably occur in nature, since they have been found repeatedly in organisms which have undergone structural alterations as a result of treatment with X-rays or other agents which promote chromosome breakage. The first one is based upon the fact that meiotic bivalents, particularly those of small chromosomes, may undergo non-disjunction at the first meiotic metaphase, so that both members of the pair are included in a single gamete. If such a gamete bearing an extra chromosome unites with a normal gamete, the resulting plant is trisomic for a small chromosome. Such trisomics are unstable, and tend to revert in later generations to their normal disomic progenitor. Under some conditions, however, particularly in progeny of crosses between different races or species, they may persist indefinitely in a population. This gives an opportunity for translocations to occur which transfer essential genetic material of a different kind to one of the three replicated chromosomes. The latter will now be perpetuated as a normal, essential chromosome of the complement (Fig. 4.8).

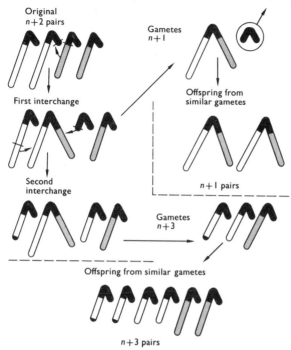

Fig. 4.8 Diagram showing how, by means of reciprocal translocation of unequal chromosomal segments, the basic chromosome number can be decreased or increased. The parts of the chromosomes coloured black are assumed to contain no genes essential for the survival of the organism. (From Stebbins.[209])

The second probable mechanism is based upon the observed fact that in progeny of heterozygotes for artificially induced translocations, plants sometimes occur which bear extra centric fragments, containing in the trisomic condition one of the normal centromeres plus adjacent chromosomal material. Such centric fragments are likewise usually lost, but may be perpetuated if essential genic material from another chromosome is transferred to them by an additional translocation.

Differences in the number and position of satellites reflect differences in the location and size of nucleolar organizer regions. These, along with differences in heterochromatization, have already been discussed in the last chapter, and so will be mentioned only incidentally here.

The concept of karyotype asymmetry

The Russian school of comparative morphology of the karyotype, led by G. Levitzky,[134] developed the concept of symmetry v. asymmetry.

A *symmetrical* karyotype is one in which the chromosomes are all of approximately the same size, and have median or submedian centromeres. Increasing *asymmetry* can occur either through the shift of centromere position from median to subterminal or terminal, or through the accumulation of differences in relative size between the chromosomes of the complement, thus making the karyotype more heterogeneous. These two tendencies are not necessarily correlated with each other, though they may be in some groups.

In order to understand the significance of karyotype asymmetry, one must look for associations between increasing asymmetry and other characteristics, both cytological ones like chromosome number and morphological or ecological characters such as the annual *v.* perennial growth habit and the presence *v.* absence of specialized morphological features. Since these other characteristics are most easily scored in terms of alternative states rather than quantitatively, comparisons between them and karyotype asymmetry are easiest to make if categories or degrees of asymmetry are also established. This can be done by recognizing three degrees of difference between the largest and the smallest chromosome of the complement, and four degrees with respect to the proportion of chromosomes which are acro- or telocentric (Fig. 4.9). This

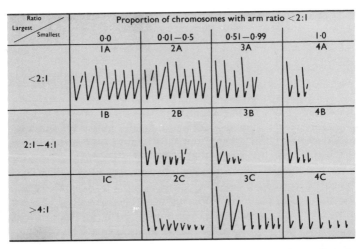

Fig. 4.9 Diagram showing a classification of karyotypes according to their degree of asymmetry. Actual karyotypes are represented, as follows: 1 A , *Aegilops mutica*; 2 A , *A. Heldreichii*; 3 A , *Crepis sibirica*; 4 A , *C. capillaris*; 2 B , *C. Mungieri*; 3 B , *C. neglecta*; 4 B , *Hypochaeris brasiliensis*; 2 C , *Muscari monstrosum*; 3 C , *Delphinium consolida*; 4 C , *Aloe zebrina*. (From Stebbins.[212])

makes possible twelve categories with respect to karyotype asymmetry, of which only ten are known to occur in higher plants. The missing two may, however, occur in animals, particularly in reptiles, many of which have karyotypes characterized by great differences in chromosome size, but with predominantly median or submedian centromeres.

Once these categories have been established and a group of species has been classified with respect to them, comparison with other characteristics can be made using the contingency chi square method. Table 4.1 shows how this method reveals an association in the Compositae, tribe Cichorieae, between greater degrees of karyotype asymmetry and lower chromosome numbers.

Table 4.1 Relation of karyotype symmetry to chromosome number in Compositae, tribe Cichorieae (except *Crepis*).

Basic number	Type of symmetry*									
	1 A	2 A	Sub total	2 B	3 A	3 B	4 A	4 B	Sub total	Grand total
$x = 9, 10$	2	24	26	2	6	2	0	0	10	36
$x = 8$	4	16	20	1	6	1	1	0	9	29
$x = 7, 6$	1	6	7	5	6	4	0	0	15	22
$x = 5, 4, 3$	0	6	6	0	8	0	1	3	12	18
Total	7	52	59	8	26	7	2	3	46	105

$x^2 = 15.94$, $n = 2$, $p < .01$.
* See Figure 4.9 for explanation.

Genera of seed plants having constant karyotypes

Karyotypes of the different species of a genus may either be much alike or may vary greatly from one species to another. Some genera of seed plants which are notable for their constancy are as follows:

Karyotype 1A: *Pinus* and other conifers; *Silene, Solanum, Microseris, Agoseris, Hieracium, Agropyron, Elymus, Hordeum, Secale, Bromus*, and many other genera of Gramineae.

Karyotype 1 or 2A: *Oenothera*.

Karyotype 2A: *Paeonia, Trillium*.

Karyotype 3B: *Lilium, Tulipa*, and related genera.

Karyotype 3C: *Delphinium, Aconitum*.

Karyotype 4C: *Aloe–Gasteria–Haworthia*; *Yucca–Agave–Furcraea*; *Hosta*.

The point must be emphasized that the similarity in karyotypes between species of these genera by no means reflects a complete similarity in chromosome structure. In many of them as, for instance, *Paeonia*,[206] *Lilium*,[184] and *Trillium*, the meiotic pairing of chromosomes in F_1 interspecific hybrids indicates that they differ from each other with respect to a large number of inversions and some translocations. Since the inversions are paracentric, and the translocations involve either relatively small or approximately equal chromosome segments, the basic features of the karyotypes have remained unaltered.

The genera listed above probably represent only a small fraction of those which have constant karyotypes. Most of the temperate genera of woody plants, such as *Quercus, Populus, Alnus, Ulmus, Acer, Viburnum, Ceanothus, Arctostaphylos,* and *Vaccinium*, probably belong in this category, as do tropical woody genera such as *Acacia, Ficus,* and *Citrus*. These genera, however, have such small chromosomes that their karyotypes are very difficult to analyse without the use of special methods, so that their classification for the present must remain uncertain.

Variation within genera with respect to the karyotype

Among the herbaceous genera which have medium-sized to large chromosomes, the great majority contain variations between species with respect to the karyotype. Four different patterns of variation can be recognized. The first is similar to that summarized in Figure 4.10 for the tribe Cichorieae, and carefully worked out in the genus *Crepis*.[6] In this genus, increasing asymmetry of the karyotype is associated with decreasing chromosome number and with increasing specialization with respect to certain morphological characteristics, such as beaked fruits and strongly modified involucral phyllaries.

A much less common pattern, in which increasing asymmetry is associated with increasing chromosome number and at least some more specialized characteristics, is found in *Clarkia*,[136] *Brodiaea* and its relatives in the Liliaceae and certain genera of Dipsacaceae (*Cephalaria–Succisa*).[64]

A third pattern is one in which two different basic chromosome numbers are associated with little difference in the symmetry of the karyotype. This pattern exists in *Anemone, Ranunculus*,[92] and *Allium*.[132] Finally, there are a few genera having a constant basic chromosome number, but variation between species with respect to karyotype symmetry. The best known of these is *Aegilops* (see Fig. 4.11).

Increasing and decreasing karyotype asymmetry

The patterns just described can be best explained by assuming, as did Levitzky, that there is a predominant trend in flowering plants toward increasing asymmetry of the karyotype. This trend is clearly

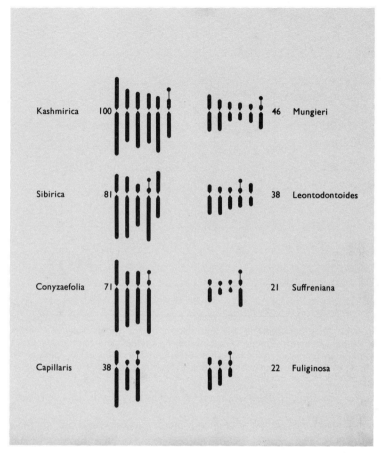

Fig. 4.10 Diagrams showing the karyotypes of eight species of the genus *Crepis* which have the largest (left) and smallest (right) chromosomes in each of the four classes of basic numbers, x = 6, 5, 4, and 3. In each case the species at the right has a greater degree of karyotype asymmetry as well as morphological specialization than the corresponding species at the left. (From Babcock.[6])

evident and has been carefully studied in *Crepis* and other genera of the Compositae, tribe Cichorieae.[212] It is recognized also in other genera of Compositae, such as *Aster*[108] and *Haplopappus*.[111] In the Ranunculaceae, tribe Helleboreae, the highly asymmetrical karyotypes found in *Delphinium* and *Aconitum* are associated with highly specialized zygomorphic flowers, while genera of this tribe having less asymmetrical karyotypes, such as

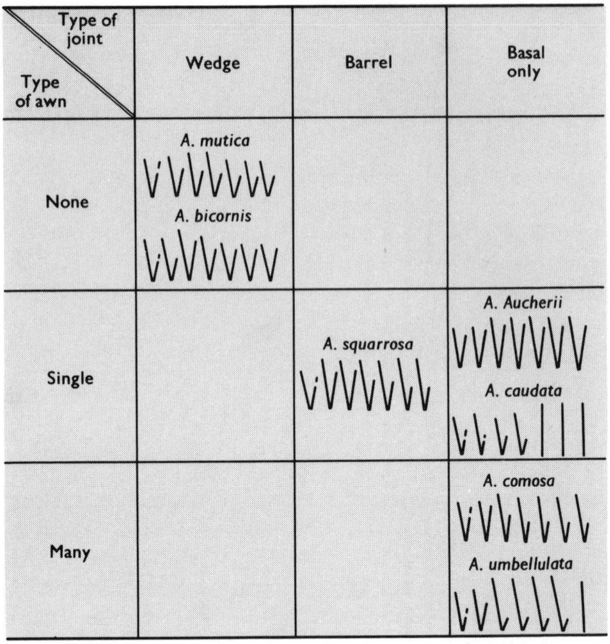

Type of joint / Type of awn	Wedge	Barrel	Basal only
None	A. mutica A. bicornis 		
Single		A. squarrosa 	A. Aucherii A. caudata
Many			A. comosa A. umbellulata

Fig. 4.11 Diagram showing the correlation in the genus *Aegilops* between increasing karyotype asymmetry and increasing specialization with respect to two morphological characteristics: the type of joint on the rachis of the spike, and the number of awns at the apex of the lemma. (From Stebbins.[212])

Caltha, Trollius, Cimicifuga, and *Nigella,* also have less specialized flowers.

The trend toward increasing karyotype asymmetry is by no means irreversible. Metacentric chromosomes may be formed through "fusion" between two acrocentrics or telocentrics in a manner similar to that shown in Figure 4.8. This process is well known and frequent in various groups of animals, particularly insects.[242] It is less common in plants, but some good examples exist. The most striking one is that of *Lycoris* (Amaryllidaceae), native to eastern Asia.[40]

Criteria for distinguishing between trends for increasing and decreasing asymmetry

Secondary trends toward decreasing asymmetry can be distinguished from the primary trends toward increasing asymmetry because the two kinds of trends are based upon different processes. Increasing asymmetry results from pericentric inversions and unequal translocations of portions of chromosome arms. It may, therefore, take place without changing the number of centromeres or of independent chromosomes. Furthermore, by converting metacentric to acrocentric chromosomes, pericentric inversions can reduce the 'fundamental number' of well developed chromosome arms.

On the other hand, 'centric fusions' between acro- or telocentric chromosomes to give metacentric chromosomes always consist of the transfer of whole arms. Consequently, they inevitably produce a reduction in the number of centromeres and chromosomes, while leaving the 'fundamental number' of arms unchanged.

We can, therefore, determine which of these processes has been chiefly responsible for chromosomal changes by comparing karyotypes of related species. In some instances, chromosome arms are all of about the same length and the number of arms is constant. The related species differ in chromosome number, and the basic number is directly proportional to the number of acro- or telocentric chromosomes in the complement. Such examples are best explained by the origin of V's from rods, and a phylogenetic decrease in chromosome number through 'centric fusions'. The most symmetrical complement, that with the smallest number of rods and the largest number of V's, is the end point of this kind of karyotype evolution.

In other examples, the karyotypes of related species differ from each other with respect to both the length and the number of chromosome arms. In these examples, the chromosome number may remain constant or nearly so while the relationships in size and form between different chromosomes may vary considerably from one karyotype to another. Such examples are best explained on the assumption of increasing karyotype asymmetry through pericentric inversions and unequal translocations. Among them the most symmetrical karyotype is the most primitive and the most asymmetrical karyotypes are derived.

Karyotype asymmetry and seed plant phylogeny

Consideration of the angiosperms alone, makes the hypothesis that the predominant trend in karyotypes has been toward increasing asymmetry the most probable one, although many reversals of this trend must have occurred. In this connection, however, the presence of highly asymmetrical karyotypes among certain gymnosperms must be considered. These are the Podocarpaceae,[96] the genera *Welwitschia*,[122] *Amentotaxus*,[30] *Ginkgo*,[131] and some species of cycads[159] (Fig. 4.12). All of these genera contain high proportions of acrocentric and/or telocentric chromosomes. Does this mean that the ancestral seed plants had highly asymmetrical karyotypes, and that the relatively symmetrical ones which exist in the

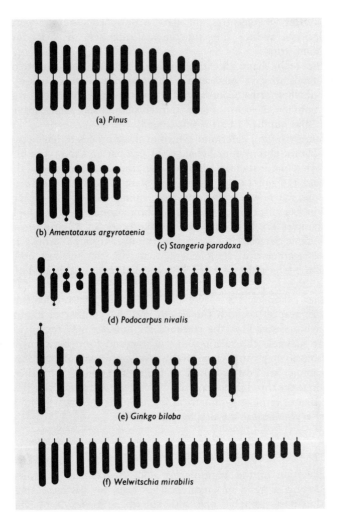

Fig. 4.12 Karyotypes of various genera of gymnosperms. **(a)** *Pinus,* showing the symmetrical karyotype characteristic of the families Pinaceae, Cupressaceae, and most genera of Taxodiaceae. **(b)**, **(c)**, Moderately asymmetrical karyotypes of *Amentotaxus argyrotaenia* (Taxaceae) and *Stangeria paradoxa* (Cycadaceae). **(d)**, **(e)**, **(f)**, Strongly asymmetrical karyotypes of *Podocarpus nivalis* (Podocarpaceae), *Ginkgo biloba* (Ginkgoaceae), and *Welwitschia mirabilis* (Welwitschiaceae). ((a) from Bowden;[21] **(b)** from Chuang and Hu;[30] **(c)** from Marchant;[159] **(d)** from Hair and Beuzenberg;[96] **(e)** from Lee;[131] **(f)** from Khoshoo and Ahuja.[122])

more primitive modern groups of angiosperms, such as Winteraceae, Illiciaceae, and Annonaceae, are derived via a trend of increasing symmetry?

The correct answer to this question depends upon recognizing clearly the difference between organisms which are truly primitive and those which, though ancient and archaic, represent ends of evolutionary trends toward increasing specialization which took place in the distant past. In the latter case, the modern organisms would be **bradytelic** according to the definition of Simpson.[194] They were highly specialized when they first evolved, but appear to us primitive and archaic because they have persisted without change for millions of years. Meanwhile, the dominant modern groups have been evolving much more rapidly, and have given rise to forms which are much more specialized along different lines. Fossil evidence for the existence of such bradytelic species is convincing in examples of animals such as *Sphenodon*, the lungfishes (Dipnoi), *Limulus*, and *Lingula*.

Of the groups of seed plants mentioned above, fossil evidence is available for the Podocarpaceae, *Amentotaxus*, and *Ginkgo*. According to this evidence, each of these groups must be regarded as a specialized end product of a long sequence of evolutionary changes. Their more primitive ancestors are all extinct, so that we can never know what their karyotypes were like. They could, however, have been relatively symmetrical. The genus *Welwitschia* has no known fossil ancestors. Nevertheless, in its enormously developed cotyledons, the only leaves which it has, and its adaptation to an extreme desert habitat, it is one of the most specialized plants in existence. The cycads, also, have a number of specialized features, particularly the great reduction in the vascular tissue of their stems.

There are, therefore, good reasons for maintaining the hypothesis that the modern gymnosperms having highly asymmetrical karyotypes are not truly primitive but archaic and specialized, and that their less specialized now extinct ancestors had relatively symmetrical karyotypes. This question can never be answered directly, but indirect evidence bearing upon it can be obtained by reviewing the karyotypes of the more primitive spore-bearing vascular plants, as well as those of bryophytes and algae. When we do so, we find that they are relatively symmetrical. Unfortunately, the more primitive spore-bearing vascular plants, the Psilotales, Lycopodiales, and Eusporangiate ferns, all have such high chromosome numbers that their karyotypes cannot easily be described.[156] Nevertheless, all of them have a high proportion of metacentric chromosomes, and there is little difference between the largest and the smallest chromosomes of the karyotype. They must, therefore, be regarded as relatively symmetrical. The bryophytes mostly have karyotypes with a fairly high degree of asymmetry,[227] but nevertheless in most of them the majority of the chromosomes are either metacentric or sub-metacentric.

Most groups of algae which have been investigated have such small chromosomes that their morphology cannot be carefully worked out. Nevertheless, those which have moderately large chromosomes have symmetrical karyotypes, as in *Oedogonium*.[105]

The most plausible hypothesis is, therefore, that in the plant kingdom as a whole, symmetrical karyotypes are usually primitive. The predominant trend is from symmetry to greater asymmetry, although reversals of this trend occur periodically. Plants which have highly asymmetrical karyotypes but belong to ancient groups are probably archaic and bradytelic rather than truly primitive.

The origin of bimodal karyotypes

Many asymmetrical karyotypes consist of two sharply distinct size classes of chromosomes, large and small. These have been termed *bimodal*.[4] The best known examples in plants are two groups of genera belonging to the Liliales. One is a group of African leaf succulents belonging to the tribe Aloinae. They have mostly red or yellow tubular flowers, adapted to bird pollination. The tribe consists of three rather large genera, *Aloe*,

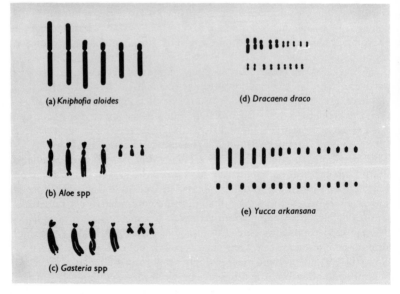

(a) *Kniphofia aloides*

(b) *Aloe* spp

(c) *Gasteria* spp

(d) *Dracaena draco*

(e) *Yucca arkansana*

Fig. 4.13 Bimodal karyotypes compared with the less asymmetrical karyotypes of related groups. ((**a**) *Kniphofia aloides,* from Webber;[239] (**b**) *Aloe* spp. and (**c**) *Gasteria* spp., from Snoad;[199] (**d**) *Dracaena draco,* from Bowden;[24] (**e**) *Yucca arkansana,* from Watkins.[238])

Gasteria and *Haworthia*, and several smaller ones. All of these species have a basic karyotype of seven pairs of chromosomes, of which four are long and acrocentric and three are short (Fig. 4.13b and c). The other is a group of xerophytes distributed chiefly in the deserts of Mexico and the southwestern United States. They are mostly large rosette plants with tough spiny leaves, although some of them in the genus *Yucca* are arboreal. The principal genera are *Yucca*, the Spanish Bayonets and Joshua Trees, and the century plants, *Agave* and *Furcraea*. All species of these genera have the basic complement ($x = 30$) of 5 pairs of large or medium-sized strongly acrocentric chromosomes and 25 pairs of very small ones (Fig. 4.13e). The mesophytic genus *Hosta* of Japan and China has a somewhat similar karyotype, and some cytologists have therefore postulated a relationship with the *Yucca–Agave* group, which is very unlikely on morphological, ecological, and distributional grounds. Recent careful studies of the *Hosta* karyotype have revealed more differences between it and *Yucca–Agave* than were previously recognized, and so have lessened the probability of a common origin.[121]

The origin of bimodal karyotypes has been explained in two ways. Darlington[40] has suggested that they are derived from symmetrical karyotypes of polyploid origin. Such karyotypes would be expected to contain many gene loci present in duplicate, so that losses of large chromosomal segments could be tolerated. The small chromosomes of bimodal karyotypes are believed to be the products of such differential loss of chromosome segments.

The second postulate is based upon the Levitzky principle of increasing karyotype asymmetry. Bimodal karyotypes could result from unequal translocation, by means of which certain chromosomes would periodically contribute segments to others of the same complement. The size of donor chromosomes would thus become reduced, and that of recipients correspondingly increased. This would give rise to karyotypes like that of the Aloinae, in which the number of small chromosomes is approximately equal to that of large ones. Karyotypes like that of *Yucca–Agave*, with their great excess of small chromosomes, could arise by an additional process involving the formation of centric fragments and the translocation to them of essential gene loci.

The best evidence for discriminating between these two hypotheses would be something which would enable one to recognize the nearest relatives and possible ancestors of the genera having strongly bimodal karyotypes. The group of Liliales which most nearly resembles the Aloinae in respect to growth habit, flowers and geographical distribution is the Kniphofinae, consisting chiefly of the genus *Kniphofia*. The species of this genus all have six pairs of chromosomes and a relatively symmetrical karyotype (Fig. 4.13d). The resemblance between the Aloinae and other

Liliales having an asymmetrical or bimodal karyotype is much more remote. Hence the bimodal karyotype of the Aloinae is best explained by increasing asymmetry and heterogeneity plus the addition of a chromosome to the complement through fixation of a centric fragment.

The *Yucca–Agave* group of genera has various morphological and ecological specializations. The century plants, as well as many species of *Yucca*, form monocarpic rosettes, which die as soon as they produce an inflorescence, after which new rosettes appear through branching of the underground parts. *Yucca* is less specialized in floral morphology than the *Agave* group, but its flowers have become peculiarly adapted to a highly specialized symbiosis with a particular genus of moths, *Tegeticola (Pronuba)*.[153] The flowers of *Agave*, having a perianth of united parts and an inferior ovary, are among the most specialized in the Liliales, and are clearly derived from some less specialized common ancestor with *Yucca*.

Plants having a growth habit similar to the less specialized species of *Yucca* together with flowers which are similarly arranged and of similar construction, but are less specialized in their method of pollination, are found in other genera placed by Hutchinson[107] in the Agavaceae: *Nolina, Dasylirion, Dracaena, Cordyline* and *Doryanthes*. The gametic numbers found in these genera are 19, 20, and 24. Their karyotypes are somewhat asymmetrical and bimodal, falling into categories 2B, 2C, and 3C of Figure 4.10 (Fig. 4.13d). Equally asymmetrical karyotypes are, however, frequent among the Liliales, and have apparently arisen many times independently of each other. On the basis of these facts, the origin of the extreme asymmetry and bimodality of the *Yucca–Agave* karyotype from one similar to *Nolina–Dasylirion–Dracaena* is best explained on the basis of pericentric inversions, unequal translocations, and the addition of modified centric fragments to provide the increase in number of small chromosomes.

Another strongly bimodal karyotype is that of the South American species of *Hypochaeris*, described on p. 107. This karyotype also is best explained on the Levitzky hypothesis of increasing asymmetry. Bimodality is, therefore, best regarded as an extreme and specialized form of karyotype asymmetry and heterogeneity, which has arisen in the same way as has asymmetry in general.

The adaptive significance of trends in the karyotype

Only two kinds of explanations can be imagined for continued, parallel trends which take place independently in unrelated groups, as do the trends toward increasing asymmetry of the karyotype. One is directed change at the gene or chromosome level, and the other is progressive natural selection of particular variants. In the case of chromosomal rearrangements, the experimental evidence is strongly against the hypothesis of internally directed trends toward greater asymmetry. Certainly, the distribution of breaks and attachments produced in chromosomes by radiations and chemical means is not at random. One of the chief deviations from randomness is, however, a tendency for segments to be removed more often from long arms and attached to shorter ones.[135] Thus, internally directed trends of karyotype variation due to spontaneous inversions and interchanges would be expected to maintain karyotype symmetry rather than promote progressive asymmetry.

The ease of separation hypothesis

Two kinds of adaptive advantages have been suggested for karyotype asymmetry. One has to do with cell dynamics: metacentric or telocentric chromosomes, as well as small chromosomes, might be able to complete the metaphase separation of their chromatids more easily than large metacentric chromosomes. Consequently, asymmetrical karyotypes might be expected to go through mitosis more rapidly than symmetrical ones, thus rendering the cells more capable of rapid proliferation. This easier separation could be based upon the fact that the final separation of chromonemata at pro-metaphase and metaphase involves proximal regions of chromatids. In acrocentric or small metacentric chromosomes, these regions are smaller than in large metacentric chromosomes, so that they could be expected to separate more easily.

There are two difficulties in this hypothesis. In the first place, variations in respect to the duration of pro-metaphase or metaphase would not be expected to serve as limiting factors determining the duration of the mitotic cycle as a whole, since these stages are relatively short compared to other stages of the mitotic cycle. Second, if cell dynamics were the principal factor controlling the increase in karyotype asymmetry, we would expect that the trend would be similar for all of the chromosomes of the complement. Asymmetrical karyotypes like those of *Lilium* and *Delphinium*, in which one or two pairs of chromosomes are large and metacentric, would be inefficient according to this concept of cell dynamics. Karyotypes would be expected to progress further toward the elimination of all large metacentrics, rather than become stabilized in an intermediate condition for millions of years, as they apparently have.

The linked gene cluster hypothesis

The alternate and more attractive hypothesis is that the chromosome arms which become longer during a trend toward asymmetry contain adaptive clusters of linked genes. Natural selection would be expected to favour translocations and inversions which would add genes to the cluster by transferring to its vicinity beneficial mutations which might arise elsewhere in the complement. In this way, any chromosome arm which might initially acquire an adaptive cluster consisting of a small number of genes would be expected to become longer by the addition of genes to the cluster which would reinforce or render more specialized the particular adaptation promoted by the cooperative action of these genes. On the other hand, chromosome arms which lacked such adaptive clusters would tend to become shorter through removal of genes from them whenever they might increase their adaptive advantage by becoming linked to the genes belonging to an adaptive cluster.

One line of evidence in favour of this hypothesis comes from the distribution of trends toward karyotype asymmetry in the Compositae, tribe Cichorieae.[212] They are not found in genera which consist chiefly of long-lived perennials inhabiting relatively stable habitats, such as *Dubyaea*, *Prenanthes*, *Microseris* subg. *Scorzonella* and certain subgenera or sections of *Lactuca*, *Sonchus*, and *Hieracium*. Furthermore, karyotype asymmetry has not evolved in species groups which, along with occupation of temporary or pioneer habitats, have become largely self fertilizing, as have species of *Microseris*, *Agoseris*, *Sonchus*, *Lactuca*, and *Hieracium*. It is also absent or poorly developed in derivatives of these genera which have become apomictic or asexually reproducing in connection with the occupation of pioneer habitats, as have *Hieracium* and *Taraxacum*. The development of karyotype asymmetry is associated with entrance into pioneer habitats, and often the evolution of annual growth cycles, along with the retention of predominant or obligate cross fertilization and the failure to evolve toward either polyploidy or apomixis. Karyotype asymmetry combined with cross fertilization and diploidy can, therefore, be regarded as a method of adaptation to pioneer, temporary habitats which is alternative to self fertilization, polyploidy, or apomixis. In species adapted to such habitats, populations must often be reduced to very small size, and new populations are often founded by a few immigrants. Under such conditions, restriction of genetic recombination permits a successful individual or small group of individuals to give rise efficiently to a much larger population which is adapted to a particular and restricted ecological niche, without loss of reproductive potential through excessive segregation. This principle, which can be called the 'infection principle' because of its resemblance to the way in which infective microorganisms achieve success, is basic to an understanding of the population dynamics of organisms which live in temporary or pioneer habitats. In higher plants, the demands imposed by this principle can be met either by the evolution of self fertilization, by polyploidy (which may be combined with self fertilization, as is explained in the next chapter), by apomixis or other forms of asexual reproduction, particularly if they are facultative and alternative with sexuality, or by increase in genetic linkage through the evolution of increasing karyotype asymmetry.

Morphological sex chromosomes in plants

The differentiation of sex chromosomes most probably involves alteration of two different chromosomal functions. In the last chapter, circumstantial evidence was presented which indicates that the initial differentiation of the sexes is associated with differential heterochromatization, and hence with differential gene action at a critical stage of development. In most dioecious species of higher plants, chromosomal differentiation has not progressed visibly beyond this stage. In nearly all higher animals, however, the karyotypes of the two sexes differ from each other with respect to both the degree of heterochromatization and the gross morphology of a particular pair of chromosomes. The number of chromosomes may also differ between the sexes. The differentiation of the sexes in animals, therefore, involves both those processes that are associated with differential gene action and those that reflect alterations of linked gene clusters.

In higher plants, relatively few examples exist of morphologically differentiated sex chromosomes,[240] and in these the degree of differential heterochromatization, although it is probably always present, has been carefully studied only in liverworts. Examples of them were given in the last chapter (see Fig. 3.3). In flowering plants, the species which have been most intensively studied are in the genera *Melandrium* (*Lychnis*) and *Rumex*.

In *Melandrium*, male plants have a conspicuous dimorphic XY pair of chromosomes at metaphase, of which the larger chromosome is the Y. Analyses of the phenotypes of male plants from which a portion of the Y chromosome has been removed[240] have provided a clear picture of how this chromosome determines the sex. Genes that suppress the development of female organs are completely linked to genes that determine the initiation and completion of anther development because they are all located on different parts of the same Y chromosome.

The situation in the genus *Rumex* is best illustrated by the example of *R. hastatulus*, an annual species native to the south-eastern United States. This species consists of three chromosomal races: two basic ones, designated Texas and North Carolina,[197] and a third, Illinois–Missouri, derived from hybrids between the first two. The Texas race has the somatic number $2n = 10$ in both sexes, and a strongly differentiated X–Y pair in male plants (Fig. 4.14). In the North Carolina race, females have $2n = 8$ chromosomes and males have $2n = 9$. At meiosis in males,

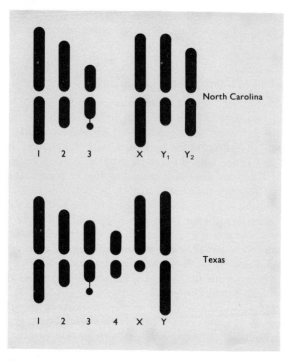

Fig. 4.14 Karyotypes of the Texas and North Carolina races of *Rumex hastatulus*, showing the gametic complement of autosomes and the somatic complement of sex chromosomes which exist in male plants. (From B. W. Smith.[197])

a chain trivalent of sex chromosomes is formed. It becomes oriented on the spindle of the first division in such a way that the X chromosome, lying in the middle of the chain, passes to one pole and the two Y chromosomes (Y_1, Y_2) pass to the other pole. The X chromosome differs from that in the Texas race, being larger and having a median rather than a subterminal centromere. The two Y's of the North Carolina race are both smaller than the single Y of the Texas race, but their combined length is greater. The smallest pair of autosomes in the Texas race is missing in the North Carolina race.

The North Carolina race has a more specialized karyotype than the Texas race, and must be regarded as derived from it. The simplest hypothesis to account for its origin is to assume the occurrence of two successive translocations, by means of which the essential chromatin of the smallest pair of autosomes in the Texas race became attached to arms of both the X and Y chromosomes, while the centromere of this

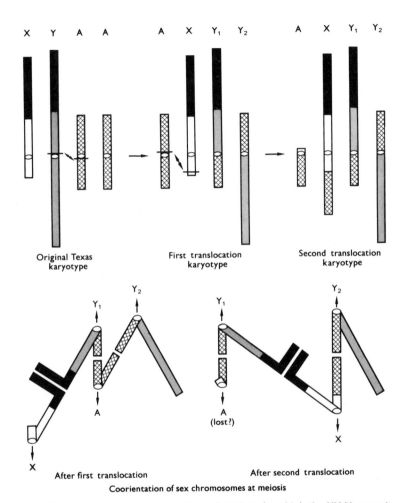

Fig. 4.15 Hypothetical sequence of translocations by which the XY₁Y₂ comple-ment of chromosomes found in the North Carolina race of *Rumex hastatulus* could have arisen from the XY complement of the Texas race, through addition of chromosomal material from the autosome 4 of Texas. Black : homologous segments of the original X and Y; white and stippled : non-homologous segments of, respectively, the original X and Y; cross hatched : segments of autosome 4 of Texas, which became translocated to all three members of the XY₁Y₂ complement of North Carolina. The centromere of North Carolina Y₂ is derived from that of Texas autosome 4. (From B. W. Smith.[197])

autosome pair became that of one of the North Carolina Y's (Fig. 4.15). The survival value of these translocations may well have been associated with their contributions to the adaptive complex of linked genes associated with sex differentiation. Similar conversions of simple to multiple sex chromosome mechanisms have been described in a number of species of insects.[243]

KARYOTYPE MORPHOLOGY AS AN AID TO TAXONOMIC CLASSIFICATION

In several instances, studies of karyotype morphology have led the way to a new and fuller understanding of the systematic relationships within a major group of plants, and to a complete reorganization of the taxonomic system of the group. The most outstanding example is the Gramineae, or grass family. In 1931, Avdulov[4] published a volume which remains today as the most thorough analysis of generic relationships within a major plant family from the cytological point of view which has ever been made. In addition to karyotypes, he studied leaf anatomy and histology, seedling development, structure of starch grains, and

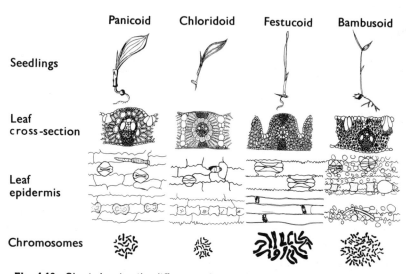

Fig. 4.16 Chart showing the differences between the largest subdivisions of the grass family with respect to diagnostic characters of seedling leaves, histology of adult leaves, and chromosomes. (From Stebbins.[211])

geographic distribution. He found that evidence from all of these characteristics pointed toward a necessary revision of the tribes of Gramineae, especially the highly heterogeneous tribes Festuceae, Agrostideae, and Phalaridae as then recognized. Avdulov's results were fully confirmed by other authors, who added further evidence to support his system from embryo morphology, vegetative anatomy, morphology of lodicules and caryopses, resistance to herbicides, and amino acid composition of seed proteins[211] (Fig. 4.16). A synopsis of the resulting system, which is now widely accepted, is presented in Figure 4.17. The tribe Festuceae as recognized by taxonomists of the nineteenth and early twentieth century has been broken up into three groups (Festuceae, Eragrosteae, Arundineae), which are distributed in three different subfamilies, while similar, though less drastic, revisions have been made on the former tribes Aveneae, Agrostideae, and Phalaridae.

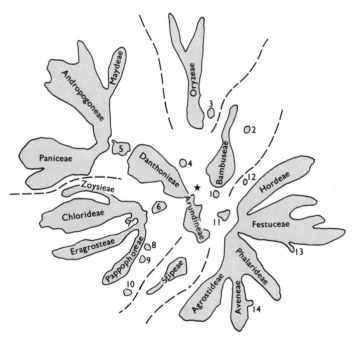

Fig. 4.17 Diagram showing the evolutionary interrelationships between the principal subfamilies and tribes of the Gramineae, based partly upon cytological and partly upon morphological characteristics. The irregular outlines represent approximately the relative size and diversity of the groups named in them, and their distance from the star in the centre is a rough indication of their degree of evolutionary specialization. The numbered circles and smaller outlines represent various genera having no close relatives.

A second example of the usefulness of karyotypes in clarifying taxonomic relationships is the family Ranunculaceae. The traditional way of sub-dividing this family has been to place into the tribe Helleboreae those genera having few carpels with more than one seed per carpel, and into the Anemoneae and Clematideae those genera having several or many carpels which are one-seeded and indehiscent. In respect to karyotype morphology, however, the family can be divided into two tribes in a completely different way.[92] One group of genera has small chromosomes with prominent heterochromatic chromocentres. This group includes *Coptis, Zanthorhiza, Hydrastis, Isopyrum* and *Aquilegia* of the Helleboreae plus *Anemonella, Thalictrum* and *Trautvetteria* of the Anemoneae. The remaining genera, which include nine assigned to the Helleboreae and five usually placed in the Anemoneae and Clematideae, have large chromosomes and irregularly distributed heterochromatic regions. The first group has principally the basic number $x = 7$, while the predominant number in the second group is $x = 8$, although $x = 7$ and $x = 6$ are also found in it.

In respect to vegetative growth habit and overall appearance of the plants, a classification based upon karyotype similarity places plants which look alike close to each other more consistently than does the classical system which is based upon characters of the carpels and fruits. For instance, *Isopyrum* and *Anemonella* are strikingly similar to each other in all characters except the number of seeds per carpel, and the similarity is almost equally great between *Caltha, Trollius,* and *Ranunculus.* This example certainly deserves further study using additional character-istics.

These examples illustrate the point that karyotype morphology can be a most useful guide to taxonomic relationships. By itself, however, it must never be regarded as of overriding importance. In some genera, such as *Colchicum,*[72] a great variety of karyotypes exists in a genus which is morphologically and ecologically very homogeneous and distinctive. This is in marked contrast to the situation just described in the Ranuncula-ceae, in which similar karyotypes exist in highly diverse and hetero-geneous groups of genera. This contrast should warn us that we can easily go astray if we use either karyology or external morphology un-critically, without careful consideration of the other characteristics of the plants in question.

KARYOTYPES, PLANT GEOGRAPHY AND ECOLOGY

One of the most vexing problems of plant geography is the determination of the place where a group originated, and the pathways over which it migrated to different parts of the earth. In vertebrate animals, some marine invertebrates, and conifers among plants, relatively direct evidence bearing upon these problems can be obtained from the fossil record. In all other groups of organisms, however, the fossil record is so scanty as to be of little use, and the problem of origins, if it is to be solved at all, must rely for its solution on indirect evidence based upon modern patterns of distribution. These distribution patterns can be highly deceptive. For instance, the generalization has been suggested that the place where a group originated is the locality where the greatest number of contemporary species of that group are located.[245] In the case of ancient groups, which have had a long and diverse evolutionary history, this generalization can be quite misleading. Take, for instance, the marsupial mammals. Their modern centre of distribution in respect to both numbers and diversity of forms is in Australia. Nevertheless, fossil evidence indicates that marsupials entered Australia from the north, since fossil marsupials are known from strata in both Asia and North America which are older than any rocks which contain them in Australia. The present diversification of marsupials in Australia is a secondary adaptive radiation, made possible by the availability of many unoccupied ecological niches, and lack of effective competition from other animals.

The example of *Hypochaeris*

An example of a similar situation in which a specialized karyotype serves as an indicator of a secondary centre of diversity is the genus *Hypochaeris*. This genus, belonging to the family Compositae, tribe Cichorieae, is defined by a group of well marked characteristics, and its boundaries can be clearly recognized. It contains more than 50 species, of which seven are native to the Mediterranean region of southern Europe, North Africa, and western Asia. Two others are found in the Alps, Caucasus, and the mountains of central Asia. A single rather distinctive species occurs in eastern Asia; there are none in North America, and the remainder of the genus, consisting of about 40 species, is in South America (Fig. 4.18). The South American species of *Hypochaeris* are very diverse in appearance, and occupy a great variety of habitats. One of them is a weed in tropical and subtropical regions of Brazil; several others are cushion plants found far above timberline in the Andes; still others are endemic to the summer-dry lomas (hills) of coastal Chile; there are species in the saline flats of the frigid, arid plains of Patagonia; and still other diverse habitats contain species of this genus.

Hypochaeris is subdivided into a number of well marked subgenera, based on the degree of specialization of the achenes and pappus. All of these subgenera are represented among the species of the Mediterranean

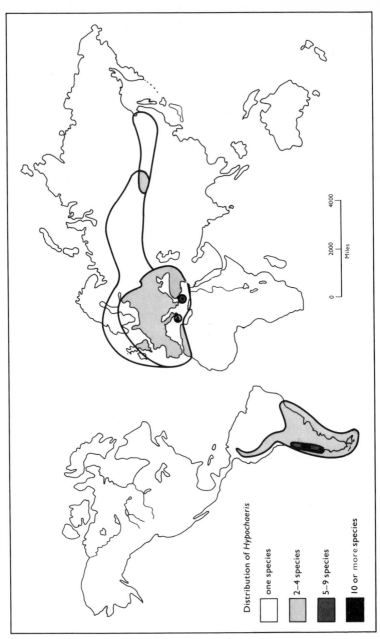

Fig. 4.18 Map showing the world-wide distribution of the genus *Hypochaeris* (Compositae, Cichorieae). Note the concentration of species in South America and the Mediterranean region, as well as its complete absence from North America.

Distribution of *Hypochaeris*

- one species
- 2—4 species
- 5—9 species
- 10 or more species

0   2000   4000
Miles

region and the Alps, while those of Asia and South America all belong to the most primitive subgenus *Achyrophorus*. The closest relative of *Hypochaeris*, *Leontodon*, is confined to western Eurasia and adjacent North Africa. In number of species, variety of habitats occupied, and cytological characteristics, the Mediterranean species of *Hypochaeris* and *Leontodon* are much alike. There are no close relatives of *Hypochaeris* in either Asia or South America.

On the basis of the facts just summarized, two interpretations of the origin and diversification of *Hypochaeris* are equally plausible. Since South America contains the largest number of species, has been the scene of the greatest amount of adaptive radiation, and contains species which belong exclusively to the most primitive subgenus, a good case could be made for assuming that the centre of origin for *Hypochaeris* is on that continent. On the other hand, the greater diversity of reproductive characteristics found among the Mediterranean species of *Hypochaeris*,

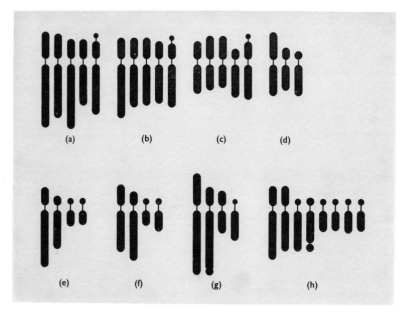

Fig. 4.19 Karyotypes of eight species of *Hypochaeris*. Top row, Old World species: **(a)** *H. grandiflora* (eastern Asia); **(b)** *H. uniflora* (western Eurasia); **(c)** *H. glabra* (native to Europe, world-wide as an introduced weed); **(d)** *H. pinnatifida* (Mediterranean region). Bottom row, South American species: **(e)** *H. brasiliensis*; **(f)** *H. foliosa*; **(g)** *H. glauca*; **(h)** *H. stenocephala,* a tetraploid. ((c) from Wulff;[249] others from Stebbins, Jenkins and Walters.[219])

plus the presence there not only of *Leontodon* but also of other related genera such as *Picris*, would argue for an origin in or near the present Mediterranean centre.

This problem is settled in a decisive manner by the karyotype morphology of the genus (Fig. 4.19). Among the western Eurasian species are found the basic numbers $x = 5, 4$, and 3, and karyotypes which differ considerably from each other, but in general are rather symmetrical, belonging to group 2A. The three alpine and Asiatic species all have $n = 5$ and similar karyotypes of group 3A. On the other hand, all of the South American species which have been investigated have the basic number $x = 4$ and a highly distinctive, strongly asymmetrical and bimodal karyotype shown in Figure 4.17. Since the investigated species include representatives of those inhabiting all of the diverse habitats in which the genus is found, it is highly probable that this remarkable karyotype occurs in all of the South American species. The probability that such an unusual and highly specialized karyotype could have arisen more than once is so low that it can be disregarded.

By far the most probable hypothesis, therefore, is that the largest and most diverse centre of variation for the modern genus *Hypochaeris* in South America is secondary. The genus probably originated in western Eurasia, along with *Leontodon*, *Picris*, and others of its relatives. The principal difficulty with this hypothesis is the problem of explaining how *Hypochaeris* got to South America, in view of its absence from North America, which would be the expected avenue of migration. This problem, however, is not peculiar to *Hypochaeris*. It exists also in *Centaurea–Centaurodendron*, *Alchemilla*, *Briza*, and a number of other genera which have centres of diversity in both western Eurasia and temperate South America, but are lacking or very poorly represented in eastern Asia and North America. In the case of *Hypochaeris*, the adaptive gene cluster hypothesis favoured in this chapter for the origin of karyotype asymmetry provides a possible solution to this problem. The assumption is made that some time during the Tertiary period *Hypochaeris* existed in North America, but only as small populations occupying pioneer habitats. These conditions would favour diversification of the karyotype, and the evolution of various asymmetrical karyotypes. A migrant from one of these small populations then entered South America, perhaps in association with larger ungulate mammals, such as camels and deer, during the Pliocene epoch. In this new environment it found rapidly expanding ecological niches, for some of which it was better adapted than were any of the native South American Compositae. Its populations, therefore, expanded rapidly, and as large populations were able to radiate into various niches by means of gene mutation and free recombination. The clusters of linked genes which had been essential in the small populations previously existing in North America, and which were responsible for the increasing asymmetry, were no longer adaptive in this new environment, and so were fragmented. Meanwhile the small North American populations of *Hypochaeris* became exposed to the revolutionary changes in climate and soil which accompanied the Pleistocene glaciations, and at the same time were facing competition from the great variety of aggressive genera of Compositae which exist on that continent. Their elimination by these unfavourable circumstances need not surprise us. Fossil evidence exists for the decimation in North America of another genus of herbs, the grass *Piptochaetium*, which was actually dominant there during the Pliocene epoch.[209]

KARYOTYPES CONSISTING OF CHROMOSOMES WITH DIFFUSE OR MULTIPLE CENTROMERIC REGIONS

As mentioned in Chapter 1, some groups of both animals and plants have chromosomes in which centromeric activity is not confined to one particular region, but is spread over a considerable part of the chromosome. Among archegoniate and seed bearing plants, this type of chromosome is rare. The only large group known to possess such chromosomes consists of the Juncaceae and Cyperaceae, which on the basis of morphological characters are now believed to be rather closely related to each other.

The genus in which this type of chromosome has been most intensively studied is *Luzula*. One of its species, *L. purpurea*, has chromosomes which are both the lowest in number and the largest in size of any which have diffuse centromeres. Since, however, *L. purpurea* is a specialized annual, and has no close relatives, it must be regarded as an unusual offshoot from the mainstream of evolution for this group of plants.

A much more common, wide-spread, and better known group of *Luzula* species is that which centres around the European *L. campestris*. These species exhibit a phenomenon which has not been recognized anywhere else in the plant kingdom: endonuclear polyploidy.[176] The species of the *L. campestris* group have either the somatic number $2n = 12$ or $2n = 24$. Those with the higher number have chromosomes half as large as those with the lower one (Fig. 4.20a, c).

This situation is believed to have come about through fragmentation of each chromosome in the set possessed by an ancestral form having $2n = 12$. This explanation is made more plausible by the occasional presence of plants having intermediate chromosome numbers and mixtures of large and small chromosomes (Fig. 4.20b). In both this and other

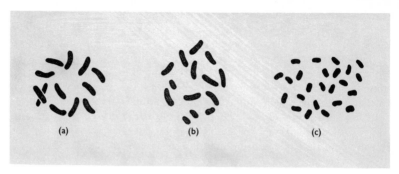

Fig. 4.20 Somatic metaphases in three different races of *Luzula spicata*, **(a)** with twelve larger chromosomes; **(b)** with ten larger and four smaller chromosomes; **(c)** with twenty-four smaller chromosomes. The total chromosomal volume is about the same in all three races. (From Nordenskiöld.[176])

subgenera of *Luzula*, true polyploidy exists, in which the species having higher multiples of the basic number $x = 6$ have chromosomes of the same size as those of the diploid species. True polyploidy apparently prevails also in the related genus *Juncus*, although the rather small chromosomes in this genus have not been much investigated.

In respect to chromosome numbers, the Cyperaceae offer a contrast to the Juncaceae, since they are characterized by extensive aneuploid series. These are best known in the genus *Carex*, and in some sections of *Eleocharis*. The species of *Carex* form an extensive aneuploid series, in which nearly every gametic number from $n = 6$ to $n = 66$ is represented.[42] The different numbers in this series are not correlated with the taxonomic subdivisions of the genus. In many instances, closely related species have different numbers of chromosomes, while the range of numbers found in one section may be similar to that found in a different, distantly related section. Several examples exist of two consecutive numbers within the same species. Hybridizations between species as well as between different populations of the same species are often heterozygous for translocations. This fact suggests that the prevalence of aneuploidy in the Cyperaceae and its comparative rarity in the Juncaceae may be due to the greater frequency of adaptive structural changes in the chromosomes in the former family as compared to the latter. As was pointed out earlier in the chapter (p. 86) centric fragments have often been reported in progenies of translocation heterozygotes. We might expect them to be particularly frequent in groups having diffuse centromeric regions, since almost any fragment formed would have part of one.

VARIATIONS IN MEIOTIC PAIRING AND THEIR
SIGNIFICANCE

An important aspect of chromosomes in relation to evolution is their
method of pairing at meiosis. When studied in individuals belonging to
stable species, chromosome pairing gives us information about the
potentialities for genetic recombination possessed by the species or
population being investigated. The nature of chromosome pairing in
interspecific hybrids may provide information concerning the role, if
any, which chromosomal changes have played in the origin of the species
concerned.

The frequency and distribution of chiasmata

Cytologists have recognized for some time that genetic crossing over
becomes visibly evident through the presence of **chiasmata** at late prophase
and first metaphase of meiosis. The formation of a chiasma can usually
be recognized by the presence of a node in a bivalent pair, but its existence
is clearly demonstrated only when the chromatids of the pairing bivalents
can be seen to exchange partners. The relationship between crossing
over and chiasma frequency is close enough so that the amount of genetic
recombination possible in a sexually reproducing, outcrossed population
can be estimated by the use of the **recombination index**, which has been
defined by Darlington[40] as the sum of the haploid number of chromosomes
and of the average chiasma frequency of all the chromosomes in a meiotic cell.

Significant differences have been found in the recombination index
between related species of some plant groups. In the Gramineae, tribe
Hordeae, for instance, the related species *Agropyron Parishii*, *Elymus
glaucus*, and *Sitanion jubatum* have progressively increasing chiasma
frequencies, the values being respectively 2.01, 2.21, and 2.79 chiasmata
per bivalent (Fig. 4.21). These values are correlated inversely with the
amount of outcrossing, which is highest in *A. Parishii*, intermediate in *E.
glaucus*, and lowest in *S. jubatum*. A similar association between low
chiasma frequencies or recombination indices in outcrossing species
and higher values in species with predominant self fertilization has been
found in *Collinsia*,[81] *Polemonium* and *Crepis*.

The best hypothesis to explain this relationship is that low recombination indices and predominant self fertilization are alternative ways of
assuring temporary reductions in the amount of genetic recombination,
thus holding together adaptive gene combinations in situations where
their retention has a particularly high adaptive value. The 'infective
principle,' mentioned earlier in this chapter, can be applied to all species
inhabiting situations in which frequent migration from one location to
another is required, and in which large populations are often built up

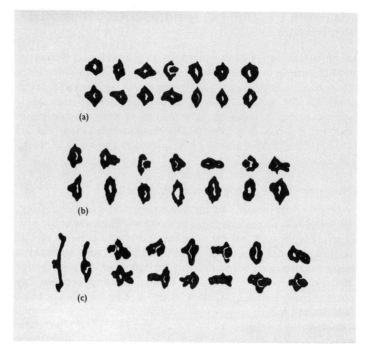

Fig. 4.21 Metaphase bivalents of (a) *Agropyron Parishii*; (b) *Elymus glaucus*, and (c) *Sitanion jubatum*, showing differences in chiasma frequency per bivalent. (From Stebbins, Valencia and Valencia.[222])

from a few founders. In such situations, an advantage is gained if the descendants of the most successful founders, which form the expanding population, retain the same gene combination which the founders possessed. On the other hand, the population retains long term flexibility, enabling it to colonize new habitats, if it is able to undergo some genetic recombination during periods of maximum expansion. Either reducing the recombination index or acquiring the capacity for predominant self fertilization are sufficient to attain this dual goal. On the other hand, if a population should acquire at the same time both a reduced recombination index and predominant self fertilization, its capacity for genetic variation might be so much reduced that it would sooner or later succumb to changes in its environment.

Another way in which chiasma formation can affect the recombination index is through localization of chiasmata to the distal regions of chromosomes, so that at metaphase bivalents appear to be attached only by the ends of their chromosomes (Fig. 4.5). This condition has been already described in the genus *Oenothera*, where it is often associated with trans-location heterozygosity and ring formation. It is well known also in *Tradescantia* and other genera. The opposite condition, localization of chiasmata near the centromere, has been found in some species of *Fritillaria*[4] and *Allium*.

Both of these kinds of chiasma localization serve the purpose not only of reducing the recombination index, but also protecting from recombination certain chromosomal regions in which chiasmata do not occur. A probable hypothesis is that the protected regions contain genes which interact in an epistatic fashion to produce characteristics having a particularly high adaptive value. Evidence for this hypothesis is, however, lacking at present.

Chromosome pairing in interspecific hybrids

As explained in Chapter 2, the pairing of homologous chromosomes at meiosis depends upon a number of factors, which have begun to be understood in a precise fashion only recently, and which are still partly unknown. Nevertheless, enough is known about certain situations which are frequently encountered so that valid interpretations of them can be made.

The most rewarding of these are examples of ring formation or chain formation in hybrids involving three or more species, in the evolution of which a succession of translocations and segmental interchange has taken place. These can be followed best in species which have sufficiently distinctive karyotypes so that each chromosome can be recognized. A good example is the analysis of certain annual species of the genus *Chaenactis* (Compositae) by Kyhos.[129] As shown in Figure 4.22, the two species having $n = 5$, *C. Fremontii* and *C. stevioides*, form F_1 hybrids having very complex chromosomal configurations at meiosis, indicating that they differ from each other by two or three translocations, depending upon the chromosomal race involved. On the other hand, in crosses with the 6-paired species *C. glabriuscula*, one race each of *C. Fremontii* and *C. stevioides* forms F_1 hybrids having a chain of only 3 chromosomes, indicating a difference with respect to a single translocation. This would suggest that both of the 5-paired species have evolved from an ancestor similar to *C. glabriuscula* by a process which included a translocation followed by the loss of a centromere, the translocations being different for each species. Additional translocations occurred in both species after they had acquired the lower chromosome number. This phylogenetic scheme is further supported by the fact that most species of *Chaenactis*, including all of the perennials, agree with *C. glabriuscula* in having $n = 6$.

In this example, as in similar ones found in *Crepis*[231] and *Haplopappus*,[112] the chromosomal races having different segmental arrangements and in

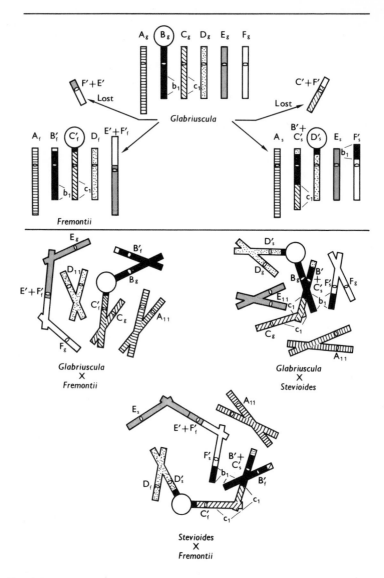

Fig. 4.22 Karyotypes of three species of *Chaenactis* (Compositae), *C. glabriuscula* (*n* = 6), *C. Fremontii* (*n* = 5), and *C. stevioides* (*n* = 5). The 5-paired species were derived by independent pathways from *C. glabriuscula,* in each case involving translocations and elimination of a part of a chromosome. The nature of these translocations was revealed by chromosome pairing in hybrids, as shown in the bottom half of the diagram. (From Kyhos.[129])

some instances different chromosome numbers have areas of distribution which are different geographically and/or ecologically. In *Clarkia* (Onagraceae), however, there are several pairs of species which differ from each other with respect to several translocations and form highly sterile F_1 hybrids, but are either completely sympatric or occupy closely adjacent distributional areas.[137,140] Moreover, in these examples, as in the others cited above, the morphological differences between the species are trivial or virtually undetectable. This situation has led Lewis to postulate that in *Clarkia* and perhaps other genera saltational speciation occurs through the simultaneous occurrence of numerous translocations, perhaps triggered off by an unusual complex of environmental factors. Although this hypothesis does appear to explain more easily than any other the facts as they are known in *Clarkia*, comparable situations have not yet been found in other genera. Consequently its general application, if possible at all, must await further cytogenetic information on other groups.

Cryptic structural differences and their significance

Only a minority of interspecific hybrids are characterized by heterozygosity for large, easily recognized chromosomal differences, which give rise to multivalent chains and rings at meiosis. Much more common are hybrids which at meiosis show reduced chromosomal pairing for no obvious reason, or which at pachytene have mismatched chromomeres or small portions of chromonemata which are irregularly paired.[86] Not uncommonly, the chromosomes may appear to be paired perfectly at first metaphase, but the hybrid is nevertheless highly sterile because the majority of the pollen grains fail to develop.

One explanation for this situation is ***cryptic structural hybridity***.[209] This hypothesis assumes that the parental species differ with respect to many small rearrangements of chromosomal segments, particularly interstitial translocations, which do not involve chromosome ends. Because of the way in which chromosomes pair at meiosis, such differences would not be expected to give rise to multivalent rings or chains, but only to a looser pairing at pachytene, a reduction in chiasma frequency, and the frequent appearance of unpaired chromosomes at first metaphase, due to failure of chiasma formation. Chromosome pairing usually begins either at the ends of the chromosomes or their centromeres and continues, zipper fashion, along the chromonemata. If the regions at which pairing is initiated are homologous, pairing can begin, and non-homologous interstitial regions, if they are not very large, are likely to be associated in spite of their lack of homology. Once this happens, segregation to the gametes of the non-homologous regions is bound to be abnormal. If an interchange of a small chromosomal segment is involved, half of the resulting gametes will be deficient for one and duplicated for the other of the interchanged segments (Fig. 4.23).

Evidence from chromosome behaviour and pollen abortion in maize, a diploid species, indicates that cryptic structural hybridity can explain the presence of pollen sterility in the absence of any chromosomal irregularities except for reduced pairing at metaphase. This evidence indicates that deficiencies for 3 to 12 per cent of the length of a chromosome are

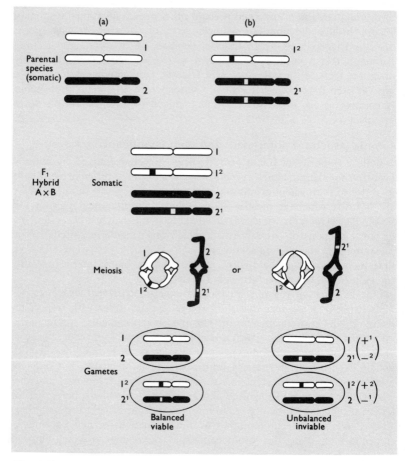

Fig. 4.23 Diagram to show how heterozygosity for an interchange involving a small interstitial segment of a chromosome can lead to good pairing of homologous chromosomes at meiosis, but consequent semi-sterility due to the presence of 50 per cent of gametes which contain deficiencies of duplication for the segments involved.

enough to render a pollen grain inviable. Interchanges involving chromo-somal segments of this length are much more likely to give rise to non-homologous pairing of the rearranged segments in otherwise normally-paired bivalents than to multivalent formation. If different interchanges of this sort are present in different regions of a chromosome, or on different bivalents, they will segregate independently of each other. When this happens, the percentage of viable pollen produced by the hybrid will be $\frac{1}{2}^n$, where n is the number of independently segregating translocated segments. Given any appreciable number of such segments, the sterility of the hybrid will be great. If $n = 7$ or more, the amount of viable pollen will be less than one per cent. The fact must be emphasized that it is the *number* of such rearranged segments, rather than their individual size, which contributes to the sterility. If such differences exist between species having the common gametic numbers $n = 6, 7, 8$, or 9, then the hybrid will be almost completely sterile if the members of each otherwise homologous pair of chromosomes differ from each other with respect to one of these small rearranged segments.

Small inverted regions are not likely to contribute in this fashion to hybrid sterility, since their genes will segregate normally unless crossing over occurs in the inverted regions, which is unlikely. Since, however, they may often pair in a non-homologous fashion and so be unable to form chiasmata, the presence of heterozygosity for many small inver-sions may reduce chiasma frequency to the extent that it affects metaphase pairing and gives rise to univalents at later stages of meiosis.

Evidence for the existence of cryptic structural differences

Three kinds of evidence exist in favour of cryptic structural hybridity. The first of these is that in certain genera having very large chromosomes, the karyotypes of related species, although they are closely similar in the overall conformation of the chromosomes, nevertheless differ with respect to recognizable details of certain chromosomal regions. In *Trillium*, these details can be revealed more clearly by exposure to low temperatures (cf. p. 62). Consequently, one can see that with respect to a single chromosome pair the difference between two species, which are separated from each other by a barrier of hybrid sterility, is a continuation of the kind of difference which exists between two interfertile races of the same species (Fig. 4.24).

Secondly, in plants having clearly analysable pachytene stages, irregu-larities of chromosome pairing, in the form of small bulges or fold-backs, or of mismatched chromomeres, can be detected in F_1 hybrids. An example is the hybrid between *Plantago insularis* and *P. ovata* (Fig. 4.25).

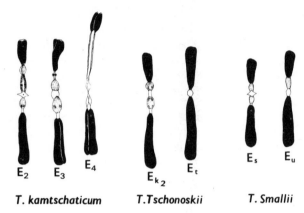

E_2 E_3 E_4 E_{k_2} E_t E_s E_u

T. kamtschaticum   **T.Tschonoskii**   **T. Smallii**

Fig. 4.24 Chromosome variability in the E chromosomes of Japanese species of
Trillium. At left (E_2, E_3, E_4) three allocyclic variants found in various populations of
T. kamtschaticum. Centre, representative E chromosomes characteristic of the
two genomes of the tetraploid species *T. Tschonoskii.* Right, the same for the
tetraploid species *T. Smallii.* (From Kurabayashi.[126])

A third kind of evidence comes from *preferential pairing* in polyploids
obtained from a sterile interspecific hybrid. This property is defined as
the greater likelihood of pairing between chromosomes which are homologous throughout their length as compared to chromosomes which
possess in common only certain homologous regions, while differing in
other regions with respect to gene content, gene arrangement or both.
Since pairing is based upon affinities between specific regions of chromosomes rather than the chromosome as a whole, we would expect that
partly homologous chromosomes would pair fairly regularly when present
together, and accompanied only by entirely non-homologous chromosomes.
On the other hand, if two completely homologous chromosomes exist
in a cell accompanied by one or two chromosomes which are only partly
homologous to them, in the majority of cells chiasma formation and
metaphase pairing will be only between completely homologous chromosomes. This will alter genetic segregation in such a way that recombination
of parental chromosomes involving deficiencies and duplications will
not occur, and the polyploidy will have converted the sterile hybrid into
a fertile polyploid. Consequently, an operational test of whether or not
cryptic structural hybridity exists in a hybrid which is highly sterile in
spite of regular bivalent formation at first metaphase is the test of chromosome doubling. If artificial doubling of the chromosome number causes

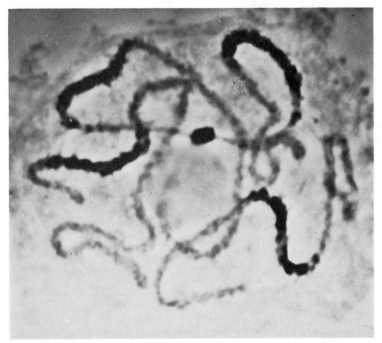

Fig. 4.25 Pairing of parental chromosomes at pachytene in the hybrid *Plantago ovata* × *insularis*. Note the irregularity which is due to small structural differences found in the heterochromatic region of Chromosome 2 (lower right). (Photo by Alva Day Whittingham.)

a sudden, dramatic increase in fertility, the existence of cryptic structural hybridity can be suspected. This result has been found in sterile hybrids with good metaphase pairing, as in *Primula verticillata* × *floribunda* and *Elymus glaucus* × *Sitanion jubatum*.[209]

A further test for the existence of preferential pairing, which may have resulted from cryptic structural hybridity, can be obtained in species having their chromosomes marked with two or more easily recognizable genes. The technique is to produce two different amphiploids between the same two species, one homozygous for recessive and the other for dominant alleles at the marker loci. These two amphiploids are then intercrossed, and the resulting F_1 is test crossed to its double recessive parent. The proportion of dominant to recessive phenotypes is recorded in the F_2 progeny. If a high degree of pollen fertility in the F_1 hybrid makes selective elimination of gametes unlikely, the frequency with which these recessives are recovered will be inversely proportional to the amount of preferential pairing in the F_1. This method has been used by Gerstel and his associates[82] to show that the sterility found in hybrids between diploid American species of the genus *Gossypium* is probably due to cryptic structural hybridity (Table 4.5).

Table 4.5 Segregation ratios in hybrid polyploids of different genomic origin in cotton (*Gossypium*). The ratios with respect of polyploids having four genomes of the A (Old World) type are relatively low; those having only two A and two D sets are high; while those involving segregation between different D genomes are highly variable. (From data of D. U. Gerstel and L. L. Phillips.[82])

Origin of polyploid	Genome formula	Leaf shape (L)	Leaf or plant colour (R)	Petal colour (Y)	Pilosity (H)
G. arboreum × herbaceum	A¹A¹A²A²	10 : 1	—	5 : 1	—
G. arboreum × Thurberi	A¹A¹DᵗDᵗ	17 : 1	158 : 0	—	—
G. hirsutum × herbaceum	A²A²AADD	5.1 : 1	5.0 : 1	5.2 : 1	4.9 : 1
G. hirsutum × Raimondii	AADDDʳDʳ	11.5 : 1	—	429 : 0	443 : 1
G. hirsutum × Thurberi	AADDDᵗDᵗ	22.4 : 1	42.7 : 1	—	—

Chromosomal and genic sterility of hybrids

In earlier reviews of the phenomenon of hybrid sterility[55,209] two kinds of sterility were recognized, genic and chromosomal. Theoretically, genic sterility was defined as due to disharmonious interaction of parental genes which gives rise to abnormal sex organs and/or abnormal meiotic behaviour independent of chromosome homology or the lack of it. Chromosomal sterility, on the other hand, was regarded as due to abnormal segregation to the gametes of either whole chromosomes or of abnormal, unbalanced combinations of chromosomal segments, produced by crossing over in bivalents made up of partly homologous chromosomes. Operationally, the proposed criterion for separating genic from chromosomal sterility was the test of polyploidy. If the initial sterility of the hybrid was largely or completely removed by chromosome doubling, the basis of the sterility was said to be chromosomal, but if the doubled hybrid remained as sterile as the undoubled one, the basis of their sterility was regarded as genic.

This distinction and its operational criterion have been largely nullified by our new understanding of epistatic gene interaction. When the definition was made, geneticists were thinking almost entirely in terms of a one-to-one relationship between a gene and a primary metabolic, enzyme-controlled process. Modern molecular genetics has shown clearly that many primary metabolic processes require harmonious interactions between genes at two or more different loci, which may even be situated on different chromosomes. The most common of these are chemical reactions which are catalysed by enzymes consisting of two or more different kinds of subunits, as is lactate dehydrogenase in vertebrate animals. Before these enzymes can become active, their two kinds of subunits must join together in a very precise fashion to form the final molecule. The fit between subunits depends upon their gene-coded primary structure. One might expect that rather often during the divergence of two species from each other, a particular enzyme would become altered in each of the descendent species in such a way that a subunit derived from one species could not unite with a subunit from the other species to form a functional enzyme.

If this should happen, functional enzymes would still be formed in the somatic cells of the F_1 hybrid, since they would contain the complete and viable gametic complement of both parents. On the other hand, independent assortment of chromosomes at meiosis would frequently produce microspores having one subunit of an enzyme derived only from one parent, and the other subunit derived only from the other parent (Fig. 4.26). Such cells would be unable to form the enzyme in question, and if the process which it catalysed was essential for gametophyte (pollen grain or pollen tube) development, partial sterility of the hybrid would be the result.

Individuals	Somatic Cells Genes	Somatic Cells Possible dimers	Gametophytic Cells Genes	Gametophytic Cells Possible dimers
P_1	$A_1A_1B_1B_1$	$A_1\text{–}B_1$	A_1B_1	$A_1\text{–}B_1$
P_2	$A_2A_2B_2B_2$	$A_2\text{–}B_2$	A_2B_2	$A_2\text{–}B_2$
F_1 Hybrid	$A_1A_2B_1B_2$	$A_1\text{–}B_1,\ A_2\text{–}B_2$ $(A_1B_2),\ (A_2B_1)$	$\frac{1}{4}A_1B_1$ $\frac{1}{4}A_2B_2$ $\frac{1}{4}A_1B_2$ $\frac{1}{4}A_2B_1$	$A_1\text{–}B_1$ $A_2\text{–}B_2$ $(A_1\text{–}B_2)$ $(A_2\text{–}B_1)$

Fig. 4.26 Diagram showing how hybrid sterility could be produced by a misfit between the two subunits of an enzyme which is a dimer consisting of subunits coded by genes at different loci. $A_1\text{–}B_1$ and $A_2\text{–}B_2$ represent functional dimers in species 1 and species 2, respectively. $(A_1\text{–}B_2)$ and $(A_2\text{–}B_1)$ represent misfit, non-functional dimers. Half of the gametes of the F_1 hybrids have non-functional dimer enzymes, so cannot develop. Hence hybrid is 50% sterile.

5

The Morphological, Physiological, and Cytogenetic Significance of Polyploidy

INTRODUCTION

The most wide-spread and distinctive cytogenetic process which has affected the evolution of higher plants has been polyploidy, the multiplication of entire chromosomal complements. Between 30 and 35 per cent of species of flowering plants, and a considerably higher percentage of ferns, possess gametic chromosome numbers which are multiples of the basic diploid number found in their genus. This fact, however, by no means indicates the amount of polyploidy which has taken place during the entire evolutionary history of vascular plants. As pointed out in the next chapter there is now good evidence to suggest that all genera or families having basic numbers of $x = 12$ or higher have been derived originally by polyploidy from groups having lower numbers, and that even the numbers $x = 10$ and $x = 11$ may often be of polyploid derivation. If this is true, then all of the modern species belonging to many prominent families, such as Magnoliaceae, Winteraceae, Lauraceae, Monimiaceae, Fagaceae, Juglandaceae, Salicaceae, Meliaceae, Ericaceae, and Oleaceae, are derivatives of evolutionary lines which at some time in their history have undergone polyploidy.

Why has polyploidy been of such overriding importance in the evolution of higher plants? The information presented in this chapter is designed to provide an answer to this question. The first kind of information needed for this purpose is about the effects of polyploidy when induced artificially in genetically balanced individuals belonging to a good species.

Morphological and physiological effects of polyploidy

These effects have long been known and are well described in a number of publications.[209] The most immediate and universal effect is an increase in cell size (Fig. 5.1). This does not always increase the size of the plant as a

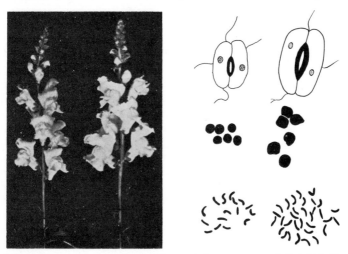

Fig. 5.1 Inflorescences, stomata, pollen and chromosomes of diploid (left) and tetraploid (right) snapdragon (*Antirrhinum majus*). (From Bamford and Winkler.[12])

whole, or even its individual organs, since a common effect of polyploidy is a reduction in the number of cell divisions which take place during development. The *gigas* effects of polyploidy are, however, commonly found, particularly in organs having a highly determinate pattern of growth, such as flowers and seeds. The increase in cell size may be reflected in larger vacuoles, and hence a higher water content of the plant as a whole, with a consequent reduction in its degree of resistance to drought and cold. This effect is, however, by no means universal.

In many instances, though not always, polyploidy causes changes in shape and texture of organs. The leaves and petals of polyploids are usually thicker and firmer than those of their diploid progenitors. Leaves and other organs are usually shorter and broader. The amount of branching is usually reduced, particularly that of tillering in polyploid grasses. The retardation of the mitotic cycle often brings about later flowering and fruiting in polyploids as compared to their diploid ancestors.

One important effect of polyploidy is a lowering of fertility and seed production. This comes about in a number of ways. In the first place, the meiosis of artificial autopolyploids is disturbed. Instead of forming exclusively bivalents, they usually form a variable percentage of quadrivalents, trivalents, and univalents. This may lead to irregularities of chromosomal segregation, and the consequent formation of gametes with unbalanced chromosome numbers. In addition, autopolyploids may be partially sterile because of various kinds of physiological unbalance, in spite of nearly regular meiotic behaviour and chromosomal segregation.

The adaptive inferiority of raw autopolyploids and ways in which it can be overcome

The characteristics just summarized by which raw, newly formed autopolyploids differ from their diploid progenitors nearly all contribute to their adaptive inferiority. It is not surprising, therefore, that such autopolyploids, which have been produced artificially in a large number of genera, have nearly always proved to be inferior to the diploid genotypes from which they arose. This inferiority is expressed in lower production of biomass per unit time, in lower seed production, and often in lowered ability to complete with diploids in artificially controlled experiments.[68] In experiments designed to test the relative success of artificially produced autotetraploids as compared to their diploid progenitors under more or less natural conditions,[210] the tetraploids have in nearly every experiment proved to be inferior. Artificially produced octoploids derived by a second doubling from experimental autotetraploids are usually sublethal.[90,133] Clearly, chromosome doubling by itself is not a help but a hindrance to the evolutionary success of higher plants.

We must assume, therefore, that in nature successful polyploidy has been accompanied by other genetic-evolutionary processes which have compensated for the initial adaptive disadvantages of raw autopolyploids. Two kinds of processes can be postulated: gradual modification of genotypes through mutation, genetic recombination and selection; and their mass modification through hybridization, either preceding or following the chromosome doubling, followed always by natural selection for adaptive segregates. Since natural selection is equally important in connection with both processes, our task is to estimate which of the two processes has been the more important: individual mutation and genetic recombination; or hybridization and the extensive genetic recombination to which it gives rise in later generations, along with stabilization of adaptive hybrid derivatives. We can make this estimate in two ways. First, we can deduce on theoretical grounds which process might be expected to be the more effective, and second, we can analyse examples of polyploidy in nature, and can make reasonable conclusions as to which processes have figured the most prominently in their evolution.

Disomic v. tetrasomic inheritance

From the theoretical point of view, the most important fact is that chromosome doubling changes the nature of genetic segregation from the disomic to the tetrasomic pattern. As pointed out in Chapter 1, tetra-somic inheritance decreases greatly the frequency of genotypes homo-zygous at a particular gene locus, and hence of those which would exhibit a characteristic controlled by a recessive allele. If the gene locus is so close to the centromere that complete linkage occurs, the F_2 phenotypic ratio for a recessive phenotype is converted from 3 : 1 (disomic) to 35 : 1 (tetrasomic). If the gene locus is so far away from the centromere that crossing over occurs regularly between them, then the F_2 ratio is 21 : 1. In any case, polyploidy greatly reduces the chance of establishment of recessive mutations. Dominant mutations are, of course, affected differently and their spread may actually be promoted by autopolyploidy.

Perhaps more important in a consideration of evolutionary processes is the effect of polyploidy on mutations which individually have small effects, and which collectively give rise to character differences that are governed by multiple factor inheritance. This importance is due to the fact that differences between natural populations with respect to adaptive characteristics are under multiple gene control. A theoretical example of what is likely to happen is presented in Figure 5.2. This figure shows the distribution of variants with respect to a quantitative character in the F_2 progeny of a cross between two parental individuals which differ with respect to genes at seven different loci. The alleles at these loci are assumed to segregate independently, to have additive effects on the character, and to be neither dominant nor recessive. If inheritance is disomic, as in a diploid, the curve of distribution is the broader, lower one expressed by the solid line. Tetrasomic inheritance, with chromosome or centromere segregation, gives the narrower, more peaked curve. This diagram shows that in crosses between races that differ with respect to quantitative characters, chromosome doubling in the progeny tends to buffer intermediate geno-types and reduce the effects of genetic segregation. Such genotypes often possess hybrid vigour. Hence this desirable characteristic may also be buffered by tetrasomic inheritance.

On the basis of these deductions, we can conclude that chromosome doubling will most often have a retarding effect on evolutionary change via mutation, genetic recombination, and selection. If this is so, we would not expect these latter processes by themselves to be very effective in counteracting the deleterious initial effects of chromosome doubling

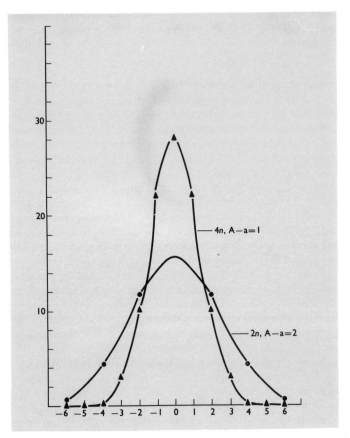

Fig. 5.2 Diagram showing the reduction in variation which would be expected in an F_2 progeny segregating for a quantitative character at the tetraploid level as compared to the diploid level. Further explanation in the text. (From Stebbins.[210])

on a balanced genotype. On the other hand, the conservatism of segregation in progeny of these doubled genotypes could be overcome by introducing through hybridization large numbers of different genes with varied effects, thus increasing the amount of genetic segregation, and enlarging the gene pool upon which selection can act.

INTERNAL FACTORS PROMOTING POLYPLOIDY

Two kinds of internal factors strongly affect the frequency of polyploidy within a group: growth habit and breeding system. In respect to growth habit, the higher percentages of polyploidy within a modern genus are found in perennial herbs, and the lowest in annuals. The figures for woody plants are intermediate, but approach more nearly those for annual than for perennial herbs (Table 5.1).

Table 5.1 Frequency of polyploidy in genera having different growth habits, in temperate and tropical regions. The table includes only genera of which ten or more species are reported, and which are homogeneous as to growth habit. Polyploids are scored on the basis of having multiples of the lowest number recorded for the genus, regardless of the relationship of this number to that of other genera in the family.

Growth habit and distribution	No. genera recorded	Per cent of genera having:			
		0–25% polyploids	26–50% polyploids	51–75% polyploids	> 75% polyploids
Woody temperate	52	60	27	11	2
Woody tropical	13	84	8	8	0
Perennial herbs temperate	145	28	37	20	15
Perennial herbs tropical	55	61	13	13	13
Annuals temperate	21	57	24	19	0

Within the category of perennial herbs there is, moreover, a correlation between efficient vegetative reproduction, particularly by rhizomes or stolons, and high percentages of polyploidy. This correlation is particularly evident in the Gramineae. In the tribe Hordeae, for example, the genus *Hordeum*, which consists entirely of caespitose bunch grasses, contains several diploid perennial species. On the other hand in the genera *Agropyron* and *Elymus*, which contain both caespitose and rhizomatous species, all of the known diploids are either caespitose or annual. All of the rhizomatous species have tetraploid, hexaploid or higher chromosome numbers. The same situation prevails in *Bromus, Festuca, Agrostis, Calamagrostis, Spartina, Phragmites, Panicum*, and other genera. In the Liliaceae, many of the strongly rhizomatous genera (*Smilacina, Clintonia, Maianthemum, Convallaria*) have basic numbers of $n = 16$ or $n = 18$, indicating a secondary polyploid origin.

These facts support the general hypothesis maintained in this discussion, that polyploids in their initial stages depend upon especially favourable combinations of circumstances for their survival and perpetuation, but that once they have become successful are more competitive and aggressive than related diploids. Most raw polyploids, particularly those having irregular meiosis must pass through a 'bottleneck' of semi-sterility. They are much better equipped to do this if they are long-lived perennials than if they are annuals, and are even more so if they can spread vegetatively by means of rhizomes or stolons. Moreover, as compared to caespitose or annual species, rhizomatous species are much more likely to crowd each other out if they adapted to similar habitats. Hence a highly successful rhizomatous polyploid has a particularly good chance of eliminating its diploid ancestor by direct competition or interaction, and thus making it extinct.

The low percentage of polyploids among most modern genera of woody plants can probably be explained by entirely different factors, which are primarily ecological and historical. This problem is discussed in the next chapter (pp. 194–196).

The most significant factor of the genetic system in relation to the frequency of polyploidy is the amount of outcrossing as compared to self fertilization. In woody plants, at least those of the temperate zone, this factor is of little consequence, since most of them are predominantly or exclusively outcrossing. In perennial herbs polyploidy appears to be just as common in self fertilizing as in outcrossing species. In annuals, however, the situation is different. Polyploidy in annual flowering plants is almost entirely confined to groups which have a high proportion of self fertilization in both the polyploids and their diploid ancestors.

This situation is most clearly evident in certain genera containing some annual species that are predominantly outcrossing and others that are largely self fertilizing. In them, the outcrossers are exclusively diploid, but the selfers may include a smaller or larger proportion of polyploids. Examples are *Eschscholzia*, *Mentzelia*, *Clarkia*, *Gilia*, *Amsinckia*, *Plantago*, and *Madia*. At first glance, one might conclude from this correlation that, in annual groups, polyploids are not likely to be of hybrid origin. This, however, is not the case. A hybrid origin has been demonstrated for annual polyploids in *Clarkia*,[139] *Gilia*,[85] *Amsinckia*[183] and *Madia*,[209] as well as for polyploids in grass genera such as *Aegilops*,[123] which do not contain any perennials or outcrossing diploids.

Consequently, the correlation between polyploidy and self fertilization among annual species is best explained as an extension of the 'bottleneck' hypothesis. If polyploidy arises in a single individual, the chances that it can produce progeny through crossing with another individual are very low, because of the hybrid incompatibility between plants having different chromosome numbers. If the plant is perennial, and particularly if it is equipped with a highly efficient mechanism for asexual reproduction, this low probability can be realized often enough so that given hundreds or thousands of years and the successive appearance of many isolated polyploid individuals, polyploids will eventually become established and successful. If, on the other hand, the initial polyploid individual is a short-lived annual, its only possibility of ever giving rise to a successful and established species lies in its capability for self fertilization, so that crossing is not required for its perpetuation.

RELATIONSHIPS BETWEEN POLYPLOIDY AND HYBRIDIZATION

These theoretical deductions lead us at once to the question: to what extent has hybridization accompanied polyploidy in the evolution of higher plants? Before we attempt to answer this question, we must be very clear in our minds as to what we mean by hybridization.

In cytological literature, the relationship between hybridization and polyploidy has often been obscured by the attempt to divide polyploids into two sharply defined categories, autopolyploids and allopolyploids. When first made by H. Kihara and his associates, this distinction was very useful in showing that when combined with wide crossing, chromosome doubling has very different effects from those which it has on balanced, non-hybrid genotypes. Subsequently, however, cytologists have attempted to set up a series of precise, rigidly defined criteria for distinguishing between the two categories, based upon external morphology, morphological similarity of metaphase chromosomes, and presence or absence of multivalents at meiosis. They have then tended to place into the category of autopolyploids all plants which do not fit the definition of allopolyploids in respect to all the criteria mentioned above, and have concluded that in the evolution of these 'autopolyploids' hybridization has not played a significant role.

The weakness of this procedure is that it implies a much too narrow concept of hybridization. By focussing attention on the effects of crossing between widely different species, which can be easily separated by the taxonomist and which have chromosomes so different from each other that they do not pair at meiosis in the F_1 hybrid, it neglects the even more important and far more common effects on polyploidy of hybridization between closely related species or between ecotypes of the same species. A more realistic estimate of these effects can be obtained by adopting and applying a much broader concept of hybridization. This is the evolutionary concept, which has been defined as follows:[213] hybridization is crossing between individuals belonging to populations which have widely different adaptive requirements. On the basis of this definition, the parents of a hybrid may be conspecific, but belong to different ecotypes. In other instances they may belong to closely related species, to widely different species, or even to different genera.

Evaluation of criteria for distinguishing the kinds of polyploids

Before we can explore the implications of this definition of hybridization for our understanding of natural polyploidy, we must evaluate critically the criteria which have been used for distinguishing between the different kinds of polyploids. A number of such criteria have been used, often uncritically, by authors of monographs as well as textbooks of cytogenetics. Too often, the assumption has been made that evolving populations will diverge from each other at equal rates with respect to a number of different, unrelated characteristics. This, however, is by no means the case. Often, evolutionary lines diverge from each other very widely with respect to morphological characteristics as well as ecological and geographic distribution, while retaining essentially similar patterns of chromosome morphology and segmental arrangement. In other instances, the external morphology of the chromosomes may remain very similar while chromosomal fine structure and gene contents diverge widely from each other.

Hybrid polyploids have been formed after crossing between populations having various degrees of divergence from each other with respect to these different characteristics. Moreover, occasional crossing between hybrid polyploids and one or both of their diploid ancestors, as well as inter-crossing between polyploids having similar but not identical hybrid origin, has often complicated greatly relationships which originally were relatively simple. The next few pages are devoted to a documentation of these generalizations.

Because of this situation, any attempt to maintain a division of natural polyploids into two discrete categories, autopolyploids and allopolyploids, is more likely to confuse than to clarify a very complex system of inter-relationships. In the present book, therefore, these two terms will be used only as ways of helping the reader to relate the present discussion of polyploids with those in other books.

The criteria which have been used are of two kinds: morphological, taxonomic characters as well as biochemical differences, all of which are the products of gene action; and cytogenetic differences, which affect directly the nature and segregation of the genes themselves.

External morphology and taxonomic key characters

The least reliable of these criteria is the assemblage of 'key characters' of external morphology which taxonomists ordinarily use in classification. The taxonomist is, rightly, most interested in ease of identification and classification. He is, therefore, reluctant to separate into different categories populations which cannot be easily differentiated by well defined characters of external morphology. These characters, however, express only a small fraction of the genetic differences between populations. Consequently, whenever wide-spread species as recognized by taxonomists are studied carefully, they are found to be highly heterogeneous genetically. This heterogeneity may be expressed partly in the form of different chromosome numbers, which may be multiples of each other. In these instances, polyploid races or 'cytotypes' with higher chromosome numbers may often contain, in addition to the set derived from the name-giving 'cytotype' of the species, other sets derived from a completely different species. Because of various kinds of gene interaction, these foreign chromosomes may not have introduced into the 'cytotype' the key morphological characters by which taxonomists define the species from which they came, or may express these characters so weakly that they are not recognized.

A good example of this situation is the grass species, *Bromus arizonicus* (Fig. 5.3). This species was first described on the basis of the usual morphological characteristics, but its close resemblance to the wide-spread *B. carinatus* caused nearly all taxonomists to place it in synonymy until its cytological characteristics became known. In the most recent taxonomic treatment of the genus,[200] it has been again relegated to the position of a 'cytotype' of *B. carinatus*.

There is, however, no doubt that half of the 84 chromosomes in the somatic cells of *B. arizonicus* have a completely different origin from any of the 56 chromosomes found in *B. carinatus*. This fact is evident both from a comparison of their karyotypes and from analysis of chromosome behaviour in the F_1 hybrid between *B. arizonicus* and *B. carinatus*. The gametic complement of *B. arizonicus* consists of 42 medium-sized chromosomes; that of *B. carinatus* contains 21 medium-sized chromosomes plus 7 much larger ones, which have no counterpart in *B. arizonicus*. In the F_1 hybrid, the 21 medium-sized chromosomes derived from *B. carinatus* are associated closely with 21 of the chromosomes derived from *B. arizonicus*, while the remaining chromosomes, both the 7 large ones derived from *B. carinatus* and the extra set of 21 derived from *B. arizonicus*, either remain as unpaired univalents, or pair only slightly with each other.

The only reasonable interpretation of this situation is that both *B. carinatus* and *B. arizonicus* are allopolyploids which share in common an ancestral species having the gametic set of 21 medium-sized chromosomes. The origin of *B. arizonicus* is from a hybrid between this common ancestral species and another one which also had 21 pairs of medium-sized chromosomes, while *B. carinatus* originated from hybridization between the same ancestor and a very different species having 7 pairs of large chromosomes. Further hybridizations have revealed the probable identity of the ancestral species concerned. The common parent probably belonged to the subgenus *Ceratochloa*, of which several 21-paired species occur in South America. The alternative 21-paired parent of *B. arizonicus* probably belonged to the subgenus *Neobromus*, while the 7-paired parent of *B. carinatus* belonged to the subgenus *Bromopsis*. Species belonging to both of these two subgenera differ from subg. *Ceratochloa* with respect to similar 'key characters',

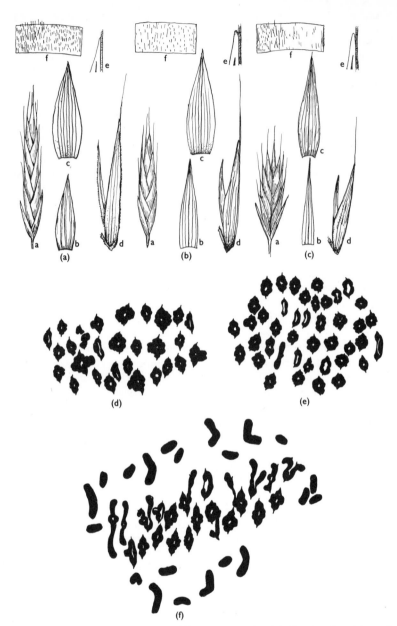

Fig. 5.3 Spikelet characteristics of *Bromus carinatus* (**a**), *B. arizonicus* (**c**), and their F₁ hybrid (**b**). Pairing of chromosomes in the two parental species (**d, e**) and the F₁ hybrid (**f**). (From Stebbins, Tobgy and Harlan.[221])

particularly lemmas rounded on the back and bearing long awns. The striking morphological differences between subg. *Bromopsis* and subg. *Neobromus*—awn bent *v.* awn straight; lemma bearing *v.* lacking prominent lateral teeth and marginal cilia—are in *B. arizonicus* so much diluted by the effects of the chromosomes derived from the *Ceratochloa* parent that they are difficult to recognize, and hence are not easily used as key characters by taxonomists.

Many similar examples can be cited: *Stipa pulchra* and *S. cernua*;[152] the *Eupatorium microstemon* aggregate,[10] and several examples in ferns. The morphological criteria used by taxonomists, although they must serve as the principal basis for distinguishing species in monographs and floras, are obviously inadequate guides to the evolutionary origin of many plant populations.

Biochemical differences

In recent years, various biochemical differences between populations have proved to be valuable aids in determining taxonomic relationships. Most frequently employed have been phenolic compounds,[1] seed proteins of an undefined nature,[115] and isozymes of particular enzymes.[190] The advantage of these differences over the conventional morphological characters is that they can usually be determined with greater objective precision than can most of the morphological differences employed, which consist of complex shapes and configurations. Moreover, in the case of proteins, the differences being studied are more closely connected with differences with respect to particular genes than are morphological differences.

At present, the weakness of biochemical criteria is that only a few kinds of compounds can be studied by those who are not trained biochemists, and in many groups significant differences with respect to these compounds are not found. In the future, one may expect that these difficulties will gradually be overcome, so that the prospects are bright for an increasing usefulness of biochemical criteria. As is described later in this chapter, they have already been useful for analysing relationships within the polyploid complex of the genus *Lotus* (p. 143 and Table 5.2, p. 136).

Chromosome morphology

The morphology of somatic chromosomes at metaphase of mitosis has often been used as a criterion for distinguishing between 'autopolyploids' and 'allopolyploids'. In a tetraploid species, the chromosomes are matched with respect to gross characteristics such as length, position of centromeres, and of satellites. If they can be grouped only into pairs, the species is regarded as allopolyploid and consequently of hybrid origin; while if they

Table 5.2 Phenolic content of diploid and tetraploid members of the *Lotus corniculatus* complex. (From P. Harney and W. F. Grant[97]).

Species	Origin	Chromosome number	D	Q	C	K	U$_1$	pC	S	F
L. corniculatus L.	France	24	+	+	+	+	−	+		−
L. corniculatus L.	France	24	+	+	+	+	+	+		+
L. corniculatus L. v. ciliatus Koch	Yugoslavia	24	+	+	+	+	+	+		+
L. c. var. *ciliatus*	Greece	24	+	+	+	+	+	+	+	+
L. alpinus Schleich.	Switzerland	12		+	+	+	+	+	+	+
L. japonicus (Regel) Larsen	Japan	12		+	+	+	+	+	+	+
L. pedunculatus Cav.	Austria	12	+	+	+	+		+·	+	+
L. pedunculatus Cav.	Morocco	12	+	+	+	+		+		+
L. pedunculatus Cav. 4n	New Zealand	24	+		+	+		+		+
L. tenuis Waldst. & Kit.	Spain	12				+				+

can be grouped into sets of four, the species is regarded as autotetraploid, and not of hybrid origin.

The fallacy of this interpretation lies in the fact that metaphase chromosomes are simply the outer shells of the genetic material, and do not reveal their contents any more than the outer aspects of two identically designed suburban tract houses reveal the internal differences in their furniture, decorations, and the people inhabiting them. This fallacy is brought out clearly by detailed studies of the karyotypes belonging to species having large, easily distinguished chromosomes. In genera such as *Lilium*, *Trillium*, *Paeonia*, and *Bromus* subg. *Bromopsis* all of the species have similar karyotypes. Hence a tetraploid derived by chromosome doubling from any diploid hybrid in these genera will have metaphase chromosomes that can be grouped into matching sets of four, and so would pass as an autotetraploid according to this criterion. Nevertheless, most of the diploid hybrids which have been made between species of these genera have irregular meiosis and are highly sterile, showing that their chromosomes are well differentiated from each other in spite of their superficial similarity. In some instances, such as *Lilium tigrinum*,[175] careful studies of the details of chromosome morphology combined with hybridizations have shown that, contrary to earlier opinions which were based upon less careful comparisons, this triploid is most probably of hybrid origin, rather than an autotriploid derivative of a single ancestral species, as was previously supposed (Fig. 5.4).

In the case of genera having large chromosomes, and particularly chromosomes in which differential, allocyclic regions can be revealed by special treatments (cf. p. 35), comparative matching of metaphase

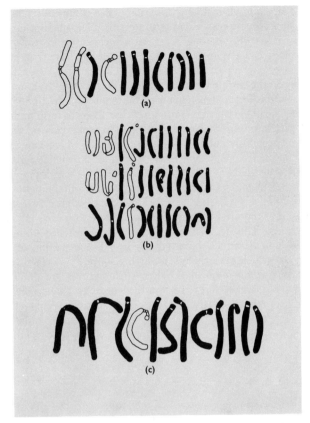

Fig. 5.4 Karyotypes of *Lilium tigrinum* and its probable progenitors. **(a)** From *L. tigrinum flaviflorum* (2n = 24), pollen grain mitosis. **(b)** From typical *L. tigrinum* (Hara, Japan), 2n = 36, root tip. **(c)** From *L. Maximowiczii,* pollen grain. (From Noda.[175])

chromosomes can provide helpful though in no case decisive information concerning the origin of a particular polyploid. If, however, the chromosomes are medium-sized or small, this method should be used with great caution or not at all.

Multivalent formation

A criterion commonly used for distinguishing between 'autopolyploids' and 'allopolyploids' is the frequency with which chromosomes associate at meiosis into quadrivalents and trivalents instead of the usual bivalent association. The difficulties with using this criterion are twofold. In the first place, even when four chromosomes are completely homologous with each other, they do not always form quadrivalents at first metaphase. Since at pachytene, chromosome segments associate only in two's, any one of four homologous chromosomes is associated with another particular homologue over only a part of its length. If chiasmata fail to form in these paired regions, the chromosomes will not remain paired at metaphase. Since chiasma frequency depends on chromosome length,[41] polyploids in plants having small chromosomes are much less likely to form multivalents than those with large chromosomes. Furthermore, since chiasma frequency is in part genotypically controlled (p. 47), diploids which contain genes for lower chiasma frequency are likely to produce polyploids forming few or no multivalents.

On the other hand, the amount of chromosomal differentiation that is sufficient to build up a barrier of hybrid sterility between two species is far less than that required to prevent chromosomes from pairing in inter-specific hybrids. This fact is evident from the existence of numerous interspecific hybrids which have good chromosome pairing at meiotic metaphase, but which nevertheless are highly sterile.[209] A classic example is *Primula verticillata–floribunda*. When such hybrids are doubled, the derived polyploid usually has fewer multivalents than a polyploid derived from one of the parental species, because of preferential pairing (cf. p. 46).

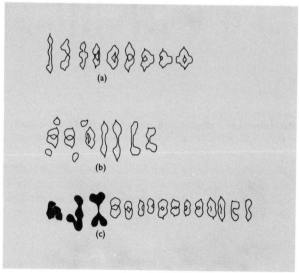

Fig. 5.5 Chromosome pairing at meiotic metaphase in *Primula verticillata* (a), *P. verticillata* × *floribunda* (b), and the hybrid polyploid, *P. kewensis* (c). (From Upcott.[236])

Nevertheless, such hybrid polyploids often form multivalents, as in *Primula kewensis* (Fig. 5.5). They are, of course, more likely to do so if the chromosomes are large, and if chiasma frequency is high in the diploid parental species. The control of this multivalent formation by special genes is discussed elsewhere (p. 47).

One can conclude from these facts that the presence or absence of multivalent configurations in a natural polyploid may provide some indication as to whether or not it is of hybrid origin, but by itself this criterion is by no means decisive. It can be used only in connection with other cytological characteristics, as mentioned above.

Tetrasomic inheritance

A criterion which has often been used in tetraploids which have been intensively investigated genetically is the presence of tetrasomic inheritance for particular genes. If such patterns of inheritance are found, the plant is judged to be autotetraploid.

This criterion is certainly valid with respect to the particular chromosomal segment on which the gene is located. One must remember, however, that many tetraploids of hybrid origin can possess some chromosomal segments in quadruplicate, and others only in duplicate.

If disomic inheritance for a character is found in a tetraploid known to be of very recent origin, this is good evidence that the plant is of hybrid origin. On the other hand, most natural polyploids are old enough so that mutations which originated after the doubling had occurred could be well established in some populations. Hence disomic inheritance in a polyploid can indicate with equal probability either that the species is of hybrid polyploid origin, or that the gene difference in question has arisen recently. Consequently, the criterion of disomic *v.* tetrasomic inheritance is not a sufficiently reliable criterion of the origin of a polyploid to be worth the large amount of labour which is required to establish it.

Experimental hybridization and chromosome doubling

The most valuable criteria for determining the nature of origin of a polyploid can be obtained by a combination of carefully planned hybridizations and doublings of the chromosome number. An ideal scheme would be as follows. First, the diploid relatives of a particular polyploid should all be studied carefully, to find out which of them resemble it the most closely with respect to morphological, and if possible, biochemical characteristics. If the polyploid is suspected to be old and long established, this screening should be world-wide, since plant groups which have existed through the Pleistocene epoch have, during that time, undergone many alterations of their patterns of geographic distribution. Since a polyploid is most likely to be successful if it is not competing directly with its diploid ancestor in the same habitat, we might expect to find often that well established polyploids would now be living in very different localities from those in which they originated, and where their diploid ancestors are still persisting.

Once these diploid relatives have been identified, they should be hybridized with each other, as well as with the polyploid under analysis. At the same time artificial polyploids should be obtained, with the use of colchicine, from both the species and their hybrids. Hybrids should then be made between the artificial polyploids and the natural one. The resulting polyploids and hybrids could be analysed and compared with respect to the criteria mentioned above: external morphology, biochemical characters, and chromosome association at metaphase. Such analysis should make possible a reasonably accurate hypothesis concerning the origin of the polyploid in question.

The importance of synthetic interpretations

The point cannot be over-emphasized that the interpretation of the origin of a polyploid, including the question of whether or not it is entirely or partly of hybrid origin, resembles all other interpretations of phylogeny. Its validity depends upon the strength of evidence derived from many different sources. Moreover, in constructing polyploid phylogenies, one cannot assume that the diploid ancestor or ancestors of a modern polyploid species still exist in their original form, unless good evidence for their existence has been obtained. Extinction or cytogenetic modification of diploid ancestors since they participated in the origin of a polyploid are likely possibilities that must always be taken into account. In the past, many erroneous interpretations of phylogeny have been made by morphologists, taxonomists and cytologists because they have fallen into one or both of the traps of reliance on only one kind of evidence and assumption that evolutionary ancestors of modern species still exist in an unmodified form. Interpretations of the nature and origin of polyploids have been no exception to this rule.

Kinds of polyploids at the primary level

In order to understand fully the facts now available about polyploidy and hybridization, we must recognize five different kinds of polyploids at the level of one cycle of chromosome doubling. This level usually involves one or more diploid ancestors and a series of tetraploids, but the same categories can often be recognized when the undoubled species are themselves of such ancient polyploid origin that they behave cytogenetically like diploids, such as the 'diploid' North American species of *Crepis* discussed on p. 176, and the diploid species of *Malus, Crataegus*, and other members of the rose family with $x = 17$.

Non-hybrid polyploids

A few tetraploids belong to monotypic or ditypic genera that have no close relatives, and consist of one diploid and one tetraploid 'cytotype' which closely resemble each other. The example of *Galax aphylla* has long been known.[209] Another example is *Achlys triphylla*[75] (Fig. 5.6). Both of these species are undoubtedly very ancient. They are members of the Arcto-Tertiary flora, which includes many groups that have evolved very little since the beginning of the Tertiary period, 60 million years ago. Both of them belong to families (Diapensiaceae, Berberidaceae) which contain many small genera that are very distinct morphologically and, like *Galax* and *Achlys*, belong to ancient, highly stable floras. These examples, therefore, support the general hypothesis that polyploidy is a stabilizing, conservative force in evolution. In spite of their great age, *Galax* and *Achlys* have been able to produce only polyploids which, morphologically and ecologically, are very much like their diploid ancestors.

Interecotypic hybrid polyploids

A number of polyploids have been recognized as products of chromosome doubling in a fertile hybrid between two different ecotypes of the same species. A good example is cocksfoot or orchard grass, *Dactylis glomerata*[223] (Fig. 5.7). The races of this species which are predominant in Eurasia north of the Mediterranean region, and which have been extensively introduced into North America and other continents, are tetraploids. In both their morphological and ecological characteristics they are intermediate between two diploids which were described as distinct species, *D. Aschersoniana* and *D. Woronowii*. The former is strongly mesophytic, and is confined to forests, chiefly in central and northern Europe, although it also occurs in the mountains of south-eastern Europe. On the other hand, *D. Woronowii* inhabits semi-arid steppe country in south-western Asia. In respect to colour and texture of the leaves, as well as several morphological characteristics of their spikelets, *D. Aschersoniana* and *D. Woronowii* are strikingly different from each other. Typical *D. glomerata* is intermediate between them with respect to these characters as well as its habitat, which is more mesic than that of *D. Woronowii* and more sunny than that of *D. Aschersoniana*. The artificial hybrid between *Aschersoniana* and *Woronowii* is fully fertile and vigorous in both the F_1 and F_2 generations, so that these two entities, from the cytogenetic point of view, must be regarded as different ecotypes of the same species. Tetraploids produced artificially from

Fig. 5.6 Geographic distribution, outline of a typical leaf, and root tip chromosomes of *Achlys japonica* (**a**) and *A. triphylla*, 2x and 4x (**b** and **c**). (From Fukuda.[75]

this hybrid closely resemble *D. glomerata* with respect to both their morpho-logical characteristics and their production of many multivalents at meiosis. In addition to typical *D. glomerata*, there are many other tetraploid subspecies of the genus in southern Europe, North Africa, and western Asia, which in the same way combine various morphological and ecological chracteristics of different diploid subspecies. The evolutionary success of intervarietal or interecotypic hybrid polyploids is often promoted by their hybrid vigour. This characteristic is buffered by the complexity of tetrasomic inheritance in their segregating progeny.

Interspecific hybrid polyploids

As has been mentioned earlier in this chapter, species that are well isolated from each other reproductively may be closely similar with respect to both the morphological appearance of their chromosomes and their structural patterns of chromosome segments. In other instances, the evolutionary divergence of species from each other involves profound repatterning of the chromosomes. Consequently, we should not expect to find that all polyploids derived from interspecific hybrids would resemble each other with respect to the morphological similarity of their chromo-somes, the frequency of multivalents at meiosis, or the proportion of genetic differences which segregate according to the pattern of tetrasomic inheritance. Three modal situations will be described. Each of them is represented by a large number of natural polyploids.

The first of these is *Lotus corniculatus*. This tetraploid, which is common in western Eurasia and has been introduced as a forage plant in many parts of the world, was analysed many years ago as an autotetraploid,[44] and was then believed to be descended only from the diploid *L. tenuis*, which also occurs in western Europe. More recently, however, the section *Corniculatae* of *Lotus* has been found to contain four or more different diploid species, which can be recognized on the basis of slight morpholo-gical differences, occupy different geographical areas or ecological habitats, have distinctly different phenolic compounds, and can be intercrossed only with great difficulty[88,89] (Fig. 5.8). Tetraploid *L. corniculatus* does not match any one of the diploid species on the basis of these characteristics, particularly the phenolic compounds (Table 5.2), and so probably contains genes derived from at least three or four of them.

The only way in which such interspecific hybrid polyploids can be distinguished from interecotypic hybrid polyploids is by identifying their diploid ancestors, and then finding out whether or not these ancestors can exchange genes freely. One must determine whether or not the diploids can easily be intercrossed, and whether or not their hybrids possess full fertility and vigour in both the F_1 and F_2 generations. This test has been

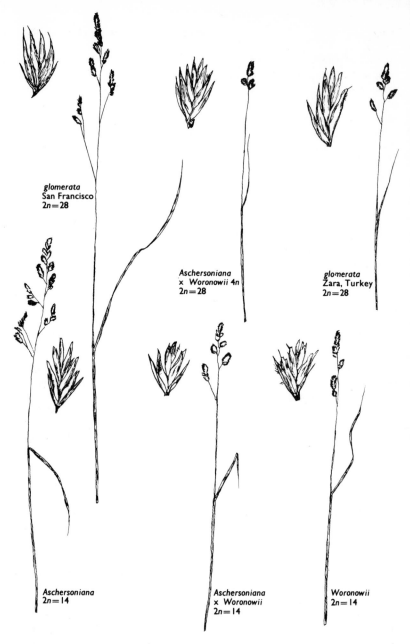

Fig. 5.7 Inflorescences and spikelets of two contrasting diploid subspecies of *Dactylis glomerata*, ssp. *Aschersoniana* and *Woronowii,* of the diploid F_1 hybrid between them, the doubled hybrid, and two tetraploid races of typical *D. glomerata.*

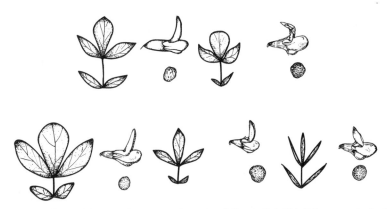

Fig. 5.8 Typical leaves, flowers and seeds of three diploid 'sibling species' of the *Lotus corniculatus* complex (lower row): *L. pedunculatus* (left); *L. Krylovii* (middle); and *L. tenuis* (right); and of two tetraploid races of *L. corniculatus* (top row). (From Zandstra and Grant.[250])

performed in only a few genera. In addition to *Lotus corniculatus*, good examples are *Zauschneria californica*[209] and tetraploid *Delphinium gypsophilum*.[138] Other probable examples are *Solanum tuberosum* and *Medicago sativa*. Well known tetraploids which are probably of hybrid origin but may be either interecotypic or interspecific, depending upon the undetermined relationships of their diploid ancestors, are *Biscutella laevigata*,[209] *Campanula rotundifolia*[17] and the tetraploid 'cytotypes' of such species as *Festuca ovina* and *Potentilla fruiticosa*.

The second modal category is intermediate with respect to chromosomal differentiation. As was pointed out in the last chapter (p. 115), many species differ from each other with respect to numerous rearrangements (inversions, translocations) of chromosomal segments, but nevertheless are exactly alike with respect to gene arrangement over a large proportion and perhaps the majority of their chromosomal complements. Hybrids between such species will have reduced pairing and irregular meiosis, and preferential pairing in tetraploids derived from such diploid hybrids will result in the appearance at meiosis either only of bivalents, or of a few multivalents plus many bivalents. Tetraploids of this nature have been designated 'segmental allopolyploids'.[209] A classical example is *Primula kewensis*. Naturally occurring examples are those which have recently evolved in eastern Washington between *Tragopogon dubius* and *T. porrifolius* as well as *T. pratensis*,[177] also *Knautia arvensis* and *Achillea collina*.[65] Perhaps the best known examples are the tetraploid and hexaploid wheats, in which bivalent formation and regular meiosis have been acquired secondarily through the action of certain genes (p. 117).

The hybrid polyploids with which cytologists are most familiar are derived from diploid hybrids between species of which the chromosomes have diverged from each other so much that little or no pairing between them is possible. Consequently the doubled hybrid forms exclusively bivalents at meiosis, and breeds true for intermediate morphological and ecological characteristics. The example of *Brassica oleracea–Raphanus sativus* has been widely cited in the literature of genetics and evolution. Other well known examples are *Galeopsis tetrahit, Nicotiana tabacum,* and the New World cottons (*Gossypium hirsutum, G. barbadense, et aff.*). Polyploids of this kind are designated in most monographs and textbooks as typical allopolyploids.

When the chromosomes of two or more ancestral species have become so strongly differentiated from each other that little or no pairing between them is possible, the gametic set of a particular diploid is inherited in derived polyploids as a single unit. Such units are termed **genomes**. In

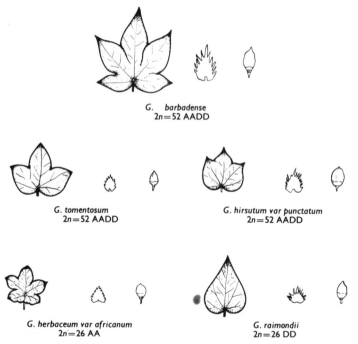

G. barbadense
2n=52 AADD

G. tomentosum
2n=52 AADD

G. hirsutum var punctatum
2n=52 AADD

G. herbaceum var africanum
2n=26 AA

G. raimondii
2n=26 DD

Fig. 5.9 Leaves, floral bracts, and capsules of three tetraploid species of cotton (*Gossypium*), and of modern relatives of their probable diploid ancestors. (From Stebbins.[214])

discussions of the phylogeny of a hybrid polyploid, the genomes are represented by a single capital letter. Thus the gametic set belonging to the Old World cultivated cottons, *Gossypium arboreum* and *G. herbaceum,* and their immediate wild relatives is designated A, and that belonging to wild species which contributed to the New World tetraploids is designated D. Hence somatic cells of the hybrid tetraploids carry the genomic formula AADD (Fig. 5.9).

Genomes are not, however, usually distinct, homogeneous categories. In many instances, two species may form a hybrid having perfect chromosome pairing at meiosis, indicating great similarity between their chromosomes. At the same time, hybrids between each of these species and a third species having an entirely different genome may differ from each other greatly in respect to chromosome pairing at meiosis.

In order to express minor differences between genomes, the letter symbols of them are often followed by modifying subscripts or superscripts. For instance, those of the different New World diploid species of *Gossypium* are usually designated D_1, D_2, etc. The fact that such modified symbols must be used emphasizes further the difficulty of maintaining a dichotomous 'either–or' kind of classification for natural polyploids.

The point must be emphasized that the modal categories just discussed are not sharply defined. Even when all facts about every natural polyploid become known, examples will exist which will be difficult to place into any one of them. They should not be regarded as completely separated compartments. Nevertheless, we can understand fully the relationship between hybridization and polyploidy only if we recognize a series of such modes.

SECONDARY MODIFICATIONS OF POLYPLOIDS

Most natural polyploids have existed for thousands or even millions of years, and have migrated from their locality of origin to many different parts of the world. During this time they have become secondarily modified and, in fact, modifications of various sorts have been essential to their success. Five different kinds of secondary modification have been important.

Mutation and genetic recombination

The theoretical reasons why mutation and gene recombination would not be expected to play as important roles in the evolution of polyploids as they do at the diploid level have already been given. Factual evidence to support this theory exists in the form of the variation patterns of many polyploid complexes. Very often, even in complexes which on the basis of phytogeographical evidence must be regarded as hundreds of thousands or even millions of years old, the range of morphological variability encompassed by all of the tetraploids is less than the total range of that found among the diploids, except for the increased sizes of parts which are the direct result of chromosome doubling.

On the other hand, diversification at a particular level of polyploidy is particularly evident in groups which have maintained themselves at this level for long periods of time. In bringing about such diversification, mutation and gene recombination must have played important roles. The best examples are those of subfamilies or families which have basic chromosome numbers of polyploid origin, like the subf. Pomoideae of the Rosaceae.[209]

Chromosomal segregation

In polyploids descended from hybrids between closely related species, chromosomes derived from different parents can segregate more or less at random. We can follow the implications of this segregation by designating the chromosomes derived from one parent $a_1, a_2 \ldots a_n$ and those derived from the other parents $a'_1, a'_2 \ldots a'_n$, where n is the basic number of chromosomes. With respect to the centromeres and the chromosomal regions immediately adjacent to them, progeny from the original doubled hybrid can have for each chromosome five possible constitutions, $a_1a_1a_1a_1$, $a_1a_1a_1a'_1$, $a_1a_1a'_1a'_1$, $a_1a'_1a'_1a'_1$, and $a'_1a'_1a'_1a'_1$. We might expect to find that natural selection would favour genotypes having a high proportion of a chromosomes in habitats similar to those occupied by the diploid a parent, and of a' chromosomes in habitats similar to those occupied by the a' parent. In this way, genetic segregation, recombination and natural selection, acting together on a hybrid autopolyploid which was advancing into a new territory, would produce a whole spectrum of genotypes and populations, encompassing the entire range of intermediacy from populations very similar to one of the original diploids in morphology and adaptiveness, through those similar to the original doubled hybrid, to populations resembling the other of the original parents. Because of the greater complexity of tetrasomic inheritance at the tetraploid level, each of these genotypes and populations would be more stable than corresponding ones at the diploid level. This stability could be further increased by the establishment of genes which would promote bivalent association and preferential pairing, thus reducing the amount of chromosomal segregation.

Just such a spectrum of morphological and ecological variants is known to exist in hybrid autopolyploids such as those of *Dactylis* and *Zauschneria*. In North America, for instance, where the diploid *Dactylis glomerata* ssp. *Aschersoniana* is absent, some forest areas, like the *Sequoia* forest of north-western California, are inhabited by tetraploid races of *D.g.* ssp. *glomerata* which in their elongate, lax leaves and spreading, open inflorescences are much like ssp. *Aschersoniana*. On the other hand, tetraploid accessions of *D.g.* ssp. *hispanica* from Palestine, Turkey, and other parts of the Middle East, where they occupy open, dry steppe country, are so similar to the diploid ssp. *Woronowii* that they can be separated from that subspecies only on the basis of their chromosome number.

Unidirectional introgression

The phenomenon of ***introgressive hybridization*** or ***introgression*** is a sequence of three processes: hybridization, back crossing, and natural selection of back cross derivatives in a habitat where they are superior to either of the original parents. When introgression takes place between a tetraploid and a diploid population, there is a strong tendency for gene flow to proceed in only one direction, from the diploid to the tetraploid. This is for two reasons. In the first place, as demonstrated in *Dactylis*, when triploids occur as sporadic hybrids in populations containing both diploids and tetraploids, progenies of these triploids from open pollination consist largely of plants having either the tetraploid number or some number approximating it.[252] Secondly, many tetraploids and diploids are so highly cross-incompatible that triploid hybrids between them cannot be formed at all. Nevertheless, many diploid species produce a small proportion of unreduced diploid gametes, through rare failures of the meiotic process. Such gametes can unite with the normal diploid gametes produced by tetraploids to give rise immediately to vigorous, fertile tetraploid hybrids. In a normal outcrossing species all of the seeds produced by such hybrids will be the result of back crosses to their tetraploid parent. Although such hybridization is undoubtedly rare, it has been recorded in *Dactylis*, *Solanum*,[162] *Grindelia*[58] and other genera. If the hybrids produced in this way, or their back cross progeny, were well adapted to a newly available ecological niche, such rare events could have evolutionary consequences far out of proportion to the rarity of their occurrence.

Evidence from variation patterns in nature suggests that unilateral introgression has played a highly significant role in increasing both the morphological range of variation and the ecological range of tolerance of many polyploids. The type of variation pattern which could have been produced most easily by this process is one in which a wide-spread tetraploid occurs sympatrically in different parts of its range with several different diploids, and in each region tends to possess races which resemble the diploids found in that particular region. This pattern is quite characteristic of *Dactylis*,[223] *Knautia*,[65] *Grindelia*,[58] *Phacelia*,[99] *Campanula rotundifolia*,[17] and many other groups.

The presence of unilateral introgression as well as chromosomal segregation both combine to make very misleading any interpretations of the origin of a polyploid which are based upon observations made entirely or chiefly in one restricted part of its geographical range. If this region is one where a related diploid is common, the races of the tetraploid inhabiting it are likely to have acquired a high proportion of genes from the associated diploid by unilateral introgression. Furthermore, polyploid races which closely resemble the sympatric diploid in their climatic preferences and probably also their morphological characteristics are likely to have a higher adaptive advantage there than elsewhere. Divergent interpretations of the origin of the same tetraploid made on the basis of material from different regions, such as those of *Anthoxanthum odoratum* by Jones[116] and by Hedberg,[100] may be explained on this basis.

Secondary hybridization

One way in which the gene pool of tetraploid species can be greatly enlarged and their evolutionary potentiality correspondingly increased is by secondary hybridization and introgression between related tetraploids. This process is particularly effective when the hybridizing tetraploids share one diploid genome in common. The common genome both increases the compatibility between the parents, as compared to that between the original diploid species, and the ease with which viable, fertile derivatives can be obtained by introgression. In addition, the diploid species which contributes the common genome may also possess some particular combination of adaptive characteristics which it transmits to all of its hybrids. It then serves as a *pivotal genome*, tying together a cluster of related tetraploids which, though usually distinct from each other even when they occur together in the same populations, can nevertheless exchange genes via introgression, particularly when hybridization takes place in disturbed habitats.

The most carefully worked out example of tetraploid species clusters are those of the grass genus *Aegilops*, studied by Daniel Zohary and his co-workers.[251] In the largest of these, the pivotal diploid parental species is *A. umbellulata*, a weedy annual found in the Middle East which possesses an unusually efficient method of seed dispersal in the form of a large number of beards or awns on its fertile scale or lemma (Fig. 5.10). Seven distinct tetraploid species contain a genome derived from *A. umbellulata*, which can be recognized both by the morphological characteristics which it introduces, as well as its distinctive karyotype (Fig. 5.10). The second genome found in these tetraploids is derived from ancestral diploids related to one of three different species complexes of *Aegilops*, those of *A. caudata*, *A. comosa*, and *A. speltoides*, which have been variously modified during later evolution at the tetraploid level. These tetraploids

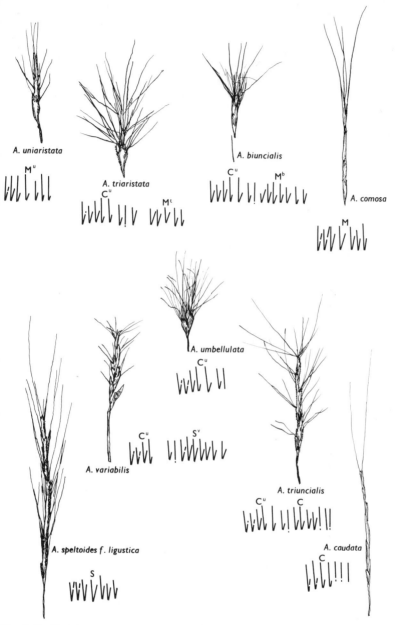

Fig. 5.10 Spikelets and karyotypes of four tetraploid species of *Aegilops* containing the pivotal genome C^u of *A. umbellulata* (centre) and of the four modern diploid species which contain counterparts of the other ancestral genomes: M^u, *A. uniaristata* (top left); M, *A. comosa* (top right); S, *A. speltoides* f. *ligustica* (bottom left); C, *A. caudata* (bottom right). The tetraploids are as follows: C^uM^t, *A. triaristata*; C^uM^b, *A. biuncialis*; C^uS^v, *A. variabilis*; C^uC, *A. triuncialis*. (Drawings of spikes made from photographs provided by J. Waines; karyotypes reproduced from figures of Chennaveeraiah.[29])

are all aggressive weeds, the most common of which have become wide-spread in the Mediterranean region. Two or more species are often found mingled together in the same habitat.

In these mixed populations hybrids are often found between two different tetraploid species. These plants can be recognized by their morphological intermediacy, by comparing them with artificial hybrids between the same two species, and by their low degree of fertility. They are completely pollen sterile, but set a small number of seeds from open pollination by the parental species. Progeny raised from such hybrids segregate widely, and recover almost complete fertility in one or two generations. Because of self pollination, which is predominant in these species as it is in most annual grasses, these fertile introgressed genotypes become fixed genetically with relative ease and, if they are of adaptive value, they may enlarge the ecological amplitude and increase the gene pool of the species concerned. The point must be emphasized here that self fertilization, although predominant in these species, is never complete, so that the introgressed genotypes, if they are interfertile with other individuals of the species concerned, may be regarded as part of its gene pool.

Secondary doubling

Occasionally, a hybrid tetraploid species may undergo a second doubling of its chromosome number to yield an octoploid, which now has two genomes, each present four times. If, according to the usual practice, the original gametic chromosome sets are designated A and B, the allotetraploid has for its somatic complement the formula AABB and the octoploid derived from it AAAABBBB. Cytogenetically, such a polyploid shows a combination of autopolyploid and allopolyploid characteristics. The best known example is the California blackberry, *Rubus ursinus*. Its unusual genomic constitution is responsible for the ease with which it can form fertile hybrids by outcrossing to distantly related species. For instance, the loganberry arose as an accidental hybrid in a garden where a female plant of this dioecious species was exposed to pollen of a diploid raspberry

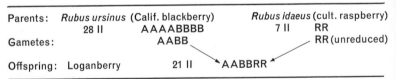

Parents: *Rubus ursinus* (Calif. blackberry)   *Rubus idaeus* (cult. raspberry)
                28 II   AAAABBBB   7 II   RR
Gametes: AABB   RR (unreduced)

Offspring: Loganberry   21 II   AABBRR

Fig. 5.11 Diagram showing how the hexaploid, fertile and stable cultivated loganberry probably originated through natural crossing between the octoploid California blackberry and an unreduced diploid pollen grain produced by the diploid raspberry.

which contained some unreduced gametes. As a result, the reduced egg of the blackberry, containing the genomes AABB, united with a raspberry pollen grain carrying RR, and the fertile hexaploid, AABBRR, was the result[209] (Fig. 5.11).

This type of secondary doubling is, however, relatively uncommon, and has not played a major role in the evolution of polyploid complexes. One reason for this is that two cycles of doubling, unaccompanied by hybridization, usually give rise to weak, abnormal plants, as in *Nicotiana*.[91]

Secondary hybrid polyploidy

A much more important secondary source of variability for polyploid complexes is hybridization followed by additional chromosome doublings. In some instances, a hybrid tetraploid is outcrossed to a third species, and the resulting triploid is doubled to produce a hexaploid which contains chromosomes derived from three original diploid species. The origin of the hexaploid bread wheats (AABBDD) from hybrids between the tetraploid emmer or macaroni wheats (AABB) and *Aegilops squarrosa* (DD) is a familiar example. In other instances, a tetraploid may be back crossed to one of its ancestral diploids, and the resulting triploid hybrid may become doubled to produce a hexaploid having twice as many chromosomes derived from one of its parents as from the other. The origin of hexaploid *A. crassa* $C^uC^uM^tM^tM^{t2}M^{t2}$ from the more wide-spread tetraploid of that species, $C^uC^uM^tM^t$, is a good example. In general, hexaploids of this kind are so similar to their ancestral tetraploids that the two are often placed in the same taxonomic species. This procedure is justifiable on the grounds that hybridization between tetraploids and closely related hexaploids can usually occur in nature when the two forms occur together and the resulting pentaploid hybrids, although they are highly sterile, can nevertheless often produce introgressed genotypes through back crosses with either of the parental species. The similarity between these forms is, therefore, based partly upon their sharing many genes originally, and partly upon subsequent gene exchange.

When secondary hybridization and polyploidy take place, the relationship between the parents of the second combination may be quite different from those which formed the first tetraploid. Thus, an interecotypic hybrid polyploid may cross secondarily with a closely related diploid, to produce a hexaploid having three genomes which are largely but not completely homologous with each other. On the other hand, it may cross with a distantly related species, to produce a bigenomic hexaploid, AAAABB. The number of possible situations is very large, and the complexity can be increased through pairing and crossing over between chromosomes belonging to partly differentiated or homoeologous genomes.

Consequently, at levels higher than tetraploidy, the 'autotetraploid' and 'allotetraploid' conditions can become so much mixed and combined with each other as to render any attempt to classify such secondary polyploids into clearly defined categories futile and meaningless.

THE POLYPLOID COMPLEX AS AN EVOLUTIONARY UNIT

A much more meaningful way of looking at the more complex products of hybridization and polyploidy is to regard them as members of a *polyploid complex*, and to study such complexes as units in themselves. Emphasis is then placed upon the processes which take place in their origin and evolution, rather than in classification and categorization of individual polyploids or groups of them. A more or less idealized diagram of a polyploid complex is presented in Figure 5.12. The theme of the next chapter will be the evolution of polyploid complexes, and its relation to ecology, plant geography, and broader questions of evolution in general.

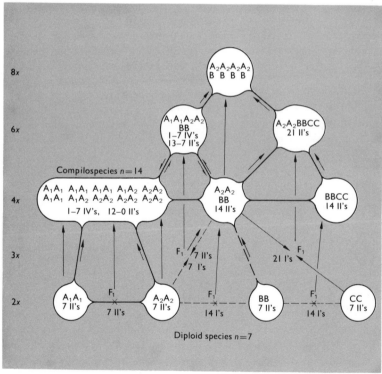

Fig. 5.12 Diagram showing a typical polyploid complex, and the various ways in which it can evolve by hybridizations and chromosome doublings.

6

Polyploidy, Plant Geography, and Major Trends of Evolution

From the material presented in the last chapter, we can conclude that polyploidy combined with hybridization has exerted a major influence on the evolution of higher plants. Its effects have been conservative. Hybridization has drastic effects on populations, since it inevitably results in the appearance of radically new gene combinations. Most of these combinations are inadaptive in any habitat. Furthermore, in a stable environment in which no ecological niches are open to colonization, all of the combinations generated by hybridization are likely to be less adaptive than those of the parental races or species, and so to be discarded by natural selection. On the other hand, when the products of hybridization are exposed to a rapidly changing environment, in which many new ecological niches are being opened up, some of these new combinations are highly likely to be better adapted to these new conditions than are any genotypes present in the old established populations. Polyploidy serves the purpose of stabilizing these valuable new genotypes, both by reducing the amount of genetic segregation, and by eliminating the sterility which exists in hybrids between well differentiated species. In addition, many individual polyploid genotypes have phenotypes which are able to tolerate a wide range of environmental conditions: they are 'general purpose genotypes'.[9] The increased size of certain organs, particularly seeds, which accompanies polyploidy may also help in the process of stabilization and establishment in new habitats, since it increases seedling vigour.

The purpose of the present chapter is to review our knowledge about the comparative distribution of diploids and related polyploids in the light of the generalizations presented in the last paragraph, and to make whatever inferences seem to be justified about the effect of polyploidy on plant evolution.

CHARACTERISTICS OF POLYPLOID COMPLEXES

Reversibility of polyploidization

Polyploidy is predominantly an irreversible trend from lower to higher levels. This irreversibility is due not to the genetic impossiblity of reversal, but to either the lowered overall adaptability or the evolutionary insignificance of its products. In the tribe Andropogoneae of the grass family, De Wet[52,53] has shown that natural autopolyploids can give rise to diploid populations. Successful reversions of this sort, however, occur only in autopolypoid populations which appear to be of relatively recent origin, and which are living sympatrically with their diploid progenitors. Such reversions do nothing more than add to the already large gene pool of the diploids, and to retard the divergent evolution of the tetraploids. Older tetraploids, which have spread beyond the range of their diploid ancestors and through the various processes already described have evolved new genotypes, invariably are either incapable of producing viable diploid offspring, or the revertants are weak, sterile, or both. This is because both mutation and the various forms of secondary hybridization and introgression tend to convert gene loci from the tetrasomic to the disomic condition. When this has happened, the diploids derived from the modified tetraploid are monosomic at many gene loci, a condition which gives rise to their weakness or sterility.

Since effective trends of polyploidy are from lower to higher levels, polyploid complexes are particularly useful for analysing problems of plant geography and phylogeny. Although they have often been used for this purpose, their possibilities in this direction have by no means been fully exploited. The relation of polyploidy to plant geography and phylogeny is a very fruitful field for further investigations.

Divergent trends in diploids and polyploids

As suggested in the last chapter, the processes of mutation and gene recombination are more effective in the evolution of diploids than of polyploids. As a result, radically new adaptive complexes, such as new floral structure, methods of pollination, and seed dispersal, are much more likely to evolve at the diploid level than in polyploids. On the other hand, the great ecological amplitude which polyploid species can acquire gives them a high degree of buffering against the environmental changes which take place over long periods of time, due to glaciations, mountain building and degradation, and overall fluctuations in the earth's climate. These factors lead to entirely different evolutionary patterns among the polyploid members as compared to the diploid representatives of any particular polyploid complex.

As the polyploid complex becomes older, its diploid members are likely to become progressively more restricted in geographic distribution and finally extinct. The only exceptions are those which evolve adaptive combinations and chromosome structures which are so radically different from those possessed by the ancestors of the polyploids that they place these newly evolved diploid populations beyond the recognizable limits of the polyploid complex.

The polyploid members, on the other hand, enlarge their gene pools and geographic distributions in the ways already described, and, as they invade the geographic areas of additional diploids, acquire genes from them. In this way they build up entities termed *compilospecies*.[54] These are wide-spread systems of polyploid populations which include chromosomes and genes derived from many different diploids, all of which are incorporated into a system within which free gene exchange is possible. The compilospecies centering about the grass species *Bothriochloa intermedia* is distributed thoughout the Old World tropics and neighbouring warm temperate regions.[54]

This differential evolution gives rise to a correlation between the age of a polyploid complex and the relative abundance of its polyploids as compared to their diploid ancestors. On the basis of this criterion we can recognize five stages of maturity: (1) initial, (2) young, (3) mature, (4) declining, (5) relictual.

Initial polyploid complexes

In their initial stages, polyploid complexes consist of one to several wide-spread diploid species within the distributional ranges of which are found one or more restricted areas occupied by polyploids. The best example is that of the three species of *Tragopogon, T. porrifolius, T. pratensis*, and *T. dubius*, described by Ownbey[177] (Fig. 6.1). In their native range in Eurasia, all three of these species are diploids having six pairs of chromosomes, and no polyploid species of *Tragopogon* are known from the Old World. In North America, *T. porrifolius* has been introduced on the west coast of the United States, from Washington to California, while *T. pratensis* is established in cool climates all across the continent. The

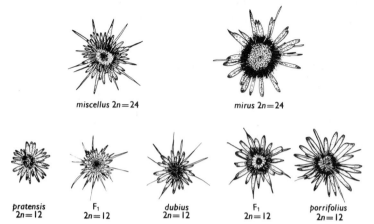

Fig. 6.1   Flowering heads of three diploid species of *Tragopogon* (bottom row); *T. pratensis* (left), *T. dubius* (centre) and *T. porrifolius* (right), and of F₁ hybrids which led to the polyploid hybrid species *T. miscellus* (top row, left) and *T. mirus* (top row, right). (From Ownbey.[177])

introduction of *T. dubius* is chiefly in the mountain states. No species of *Tragopogon* is native to the New World.

Allopolyploids between *T. porrifolius* and *T. dubius*, as well as *T. pratensis* and *T. dubius*, were first found in south-eastern Washington, in 1949, where the parental species had been introduced for only about 25 to 30 years. The exact age of the polyploids is unknown, but their recent spread is being carefully followed (Fig. 6.2).

Young polyploid complexes

The best example of a young polyploid complex is that of *Aegilops*, described in the last chapter (p. 150). The pivotal diploid species of the Cᵘ cluster, *A. umbellulata*, is still wide-spread both as a weed and in semi-natural habitats. Its range overlaps with those of the three other diploid species which contributed genomes to the tetraploids, *A. caudata*, *A. comosa*, and *A. uniaristata*, so that some degree of sympatry exists between all of its species, both diploid and tetraploid (Fig. 6.3). On the other hand, the tetraploid species *A. triaristata* and *A. triuncialis* have spread far beyond the limits of all of the diploids, and are far more aggressive and weedy. Since they occur exclusively in habitats which have been greatly modified by human activity, the most logical conclusion is that this complex began its evolution together with the beginnings of agriculture and the grazing of domestic animals, about 10 000 years ago.

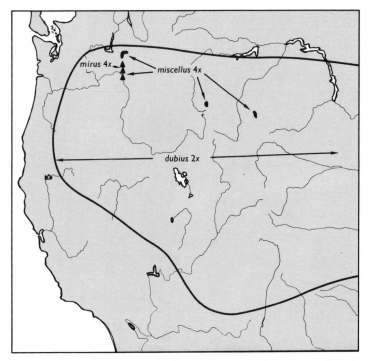

Fig. 6.2 Occurrence, generalized, of the European diploid species *Tragopogon dubius* as a naturalized weed in the western United States, and localities where its hybrid polyploids with *T. pratensis* (*T. miscellus*) and *T. porrifolius* (*T. mirus*) have been found. The diploid species *T. pratensis* is extensively naturalized throughout the area of *T. dubius*, while *T. porrifolius* occurs chiefly in the western part of the area. (From unpublished data of Marion Ownbey.)

The complex of *Lotus corniculatus*, described in the last chapter, can be regarded as intermediate between youthful and mature. Its diploid representatives are still numerous and wide-spread, although they are rarely sympatric with each other, and have diverged genetically to such an extent that diploid hybrids can be made only by artificial means. The total natural range of all of the diploids still exceeds that of the tetraploids. Nevertheless, in most of western Europe, particularly in those regions which were covered or much altered by the Pleistocene glaciation, the tetraploid *L. corniculatus* is by far the most common, dominant member of the complex.

Mature polyploid complexes

This category includes the great majority of described polyploid

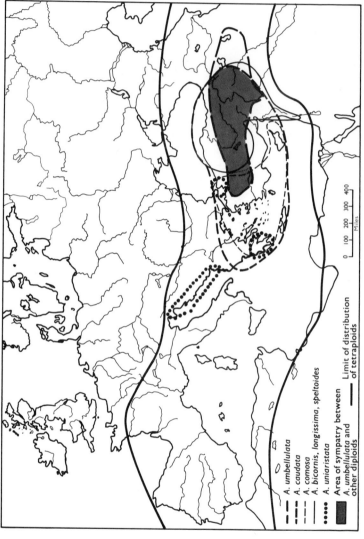

Fig. 6.3 Map of the Mediterranean region, showing the distribution of the diploid species which have entered into the *Aegilops umbellulata* polyploid complex, and the limits of distribution of the tetraploid members of this complex. (Data from Eig.[67])

A. umbellulata
A. caudata
A. comosa
A. bicornis, longissima, speltoides
A. uniaristata
Area of sympatry between   Limit of distribution
A. umbellulata and   of tetraploids
other diploids

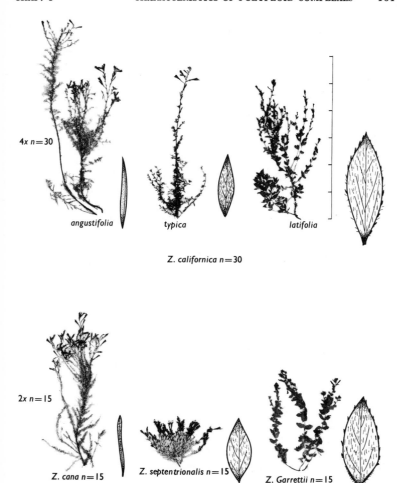

Fig. 6.4 Outlines of the growth habit and leaves of the three diploid representatives of the *Zauschneria* polyploid complex (Onagraceae), *Z. cana, Z. Garrettii,* and *Z. septentrionalis,* and of three subspecies of the tetraploid *Z. californica.* Hybrids between the diploids are vigorous and fertile, but their progeny degenerate in the F₂ and later generations. The tetraploids are freely able to exchange genes with each other. (From Clausen, Keck, and Hiesey.[33])

complexes. In them, both the morphological and ecological extremes are usually represented by diploids. These are, however, much less extensively developed than the polyploids. Their geographic ranges are more restricted, and the amount of genetic variation within their populations is usually

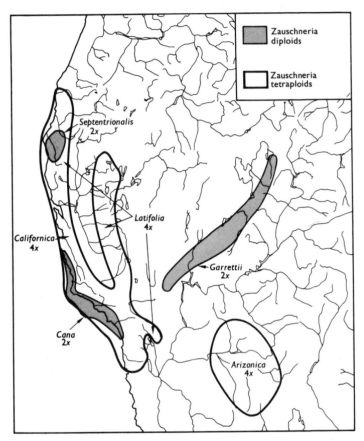

Fig. 6.5 Map of the western United States, showing the distribution of the diploids and tetraploids of the *Zauschneria* polyploid complex.

less. Furthermore, their diploids are usually allopatric with each other or sympatric in only restricted areas, so that hybridization and doubling involving different diploids has become impossible.

Several examples of mature polyploid complexes have already been described. The genera *Zauschneria* (Figs. 6.4 and 6.5) and *Dactylis* are representative of complexes which with respect to the chromosome numbers involved are relatively simple. Table 6.1 lists several other examples. In the great majority of mature polyploid complexes, the most

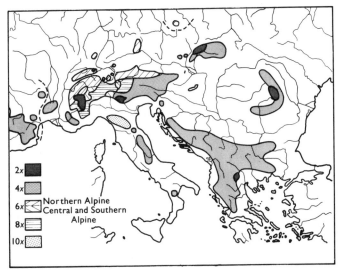

Fig. 6.6 Map showing the distribution of the polyploid complex of *Galium anisophyllum* in central and southern Europe. The diploids occur in widely separated portions of the Alps, the mountains of Central Europe and the Balkans, in which the effects of Pleistocene glaciation were minimal. The tetraploids are more wide-spread than other chromosomal levels. Hexaploids and octoploids occur chiefly in the glaciated portions of the Alps and Pyrenees. Finally a decaploid subspecies of *Galium anisophyllum* occurs in the northern Apennines of Italy. (From Ehrendorfer.[65])

wide-spread level of polyploidy is tetraploid. Hexaploids and octoploids, when present, are both fewer in number and more restricted in distribution. There are, however, some exceptions to this generalization. One of them is in the genus *Galium*. In Europe, the complex of *G. anisophyllum* consists of diploids, tetraploids, hexaploids, octoploids, and decaploids, of which the tetraploids are the commonest and most wide-spread (Fig. 6.6).

Most of the polyploid complexes in which levels higher than tetraploidy are predominant appear to have gone through two or more cycles of polyploidy, with periods of diversification and differentiation at the diploid or tetraploid level before the higher polyploids appeared. They are discussed in a later section of this chapter (p. 193).

An aneuploid complex: Claytonia *in eastern North America*

One of the most extensive series of chromosome numbers found in any groups of closely interrelated populations exists in the spring beauties,

Table 6.1 Representative sexual polyploid complexes. Y—Youthful complexes; abundance and diversity of diploids nearly equalling to exceeding tetraploids. M—Mature complexes; diploids less wide-spread than tetraploids.

Name of Complex and Reference	Basic number	Levels of polyploidy	Geographic Distribution Diploids	Polyploids	Occurrences of sympatry Between different diploids	Between diploids and polyploids
Artemisia tridentata (Y) Ward,[237] Beetle,[14] Taylor *et al.*[228]	x = 9	4x, 6x, 8x	Western North America	Similar	Wide-spread	Wide-spread
Galium multiflorum (Y) Ehrendorfer,[63] Dempster and Ehrendorfer[50]	x = 11	4x, 6x	Deserts and mountains of western U.S.	Siskiyou Mountains, Sierra Nevada, southern Great Basin	Southern part of range	Similar
Gayophytum spp. (Y) Lewis and Szweykowski[141]	x = 7	4x	Western U.S., temperate South America	Western North America, temperate South America	Throughout range of diploids	Similar
Tradescantia virginica (Y) Anderson and Sax,[3] Anderson,[2] Dean[48]	x = 6	4x	Central and southern U.S.	Rocky Mountains–Atlantic, south to northern Mexico	Central Texas	Throughout range of diploids
Achillea millefolium (M) Clausen, Keck and Hiesey,[34] Ehrendorfer,[60] Schneider[191]	x = 9	4x, 6x, 8x	Central and southern Europe, south-west Asia	Eurasia, North America	Central Europe	Similar

Species						
Campanula rotundifolia (M) Guinochet,[94] Bøcher,[18] Hubac,[106] Gadella[79]	x = 17	4x, 6x	Alps, Pyrenees, central and northeastern Europe, Arctic regions	Eurasia, North America	Not known	Alps, Pyrenees, Greenland
Clarkia rhomboidea (M) Mosquin[168]	x = 7, 5	4x (2x, 2x)	Mountains of northern and central California	Western North America	None	Throughout range of diploids
Drosera anglica (M) Wood[247]	x = 10	4x	Holarctic	Similar, plus Hawaii	Northeastern North America	Wide-spread
Fragaria spp. (M) Dogadkina,[56] Staudt[203,204,205]	x = 7	4x, 6x, 8x	Eurasia, North America	Eurasia, North and South America	Western Eurasia	Throughout most of diploid range
Gilia sect. Arachnion (M) Grant and Grant,[84] Day[45]	x = 9	4x	Deserts of south-western U.S.	Similar	None between progenitors of tetraploids	Portions of range of diploids
Phacelia magellanica (M) Heckard[98]	x = 11	4x	Western U. S.	Western North America, South America	Northern California	Through most of diploid range
Rumex paucifolius (M) Löve and Sarkar,[151] Smith[198]	x = 7	4x	Southern Sierra Nevada, north central Rocky Mountains	Mountains of western U.S.	None	Through most of diploid range
Sanicula crassicaulis (M) Bell[15]	x = 8	4x, 6x	Coastal and southern California	Pacific North America, South America	South central California	Throughout diploid range

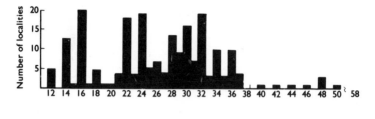

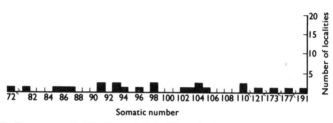

Somatic number

Fig. 6.7 Frequency distribution of the different chromosome numbers found within the single species *Claytonia virginica.* Owing to the fact that the central and southern portions of the range of the species have been sampled much more intensively than the northern and north-eastern regions, the sample may not be completely representative, but it is unlikely that many additional numbers will be found. The reproductive biology and seed fertility of the different chromosomal races have not been recorded. (From Lewis, Oliver and Suda.[142])

genus *Claytonia*, of eastern North America. In *C. virginica* there are diploids with $n = 6$, 7, and 8, which occur in localized portions of the south-eastern United States, in Tennessee, Alabama, Missouri, Arkansas, and Texas[142] (Fig. 6.7). The related species *C. caroliniana*, found principally in the Appalachian region and northward into New York, New England, southern Canada, and the Great Lakes region, consists chiefly of diploids with $n = 8$. In both species, polyploid cytotypes have been reported, but the series is by far the most extensive in forms referred to *C. virginica* (Fig. 6.8). The maximum number, $2n = \pm 191$,[188] must be regarded as the 24-ploid condition.

In contrast to most polyploid complexes, aneuploidy is general within *Claytonia*. Two reasons can be given for this anomaly. In the first place, the original diploids may once have been more wide-spread than they are now, and therefore could have occurred sympatrically and hybridized with each other. From different combinations between them tetraploids could have appeared having every number from $n = 12$ to $n = 16$, and hexaploids having every number from $n = 18$ to $n = 27$. Further inter-crossings between races and subsequent doublings could have built up the higher series of aneuploid numbers. Secondly, because of the high

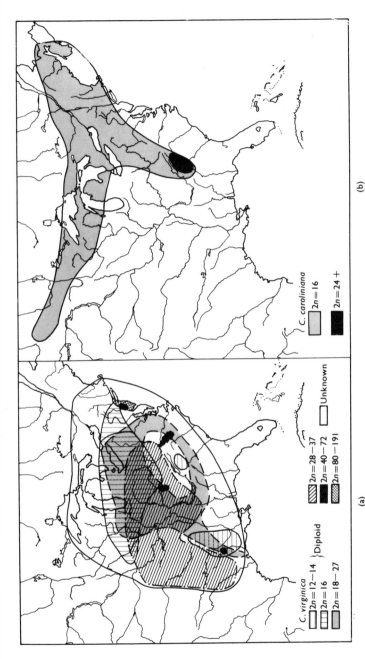

Fig. 6.8 Maps showing (a) the distribution of chromosomal races of *Claytonia virginica*, and (b) the generalized distribution of *C. caroliniana*. Note that diploids occur in three disjunct regions: the south-eastern Mississippi Valley ($2n = 14$); the southern Appalachian Mountains ($2n = 12, 14$), and probably through most of the northern half of the range of the species ($2n = 16$). The area of high chromosome numbers ($2n = 80$–191) may be more extensive than is indicated, since no counts are recorded from the regions to the north and east of New York City. (From Lewis, Oliver and Suda.[142])

degree of chromosomal duplication found among them, losses and gains of individual chromosome pairs would have less deleterious effects than in most other polyploid complexes.

Although the series found within the two taxonomic species, *C. caroliniana* and *C. virginica*, have been regarded as separate and independent 'autopolyploid' series, this supposition is not well supported by the morphological pattern of variation. The only visible differences between these species are in leaf width and the character of the inflorescence bract, the latter character being particularly unreliable. In respect to leaf width, a complete transitional series can be found in the southern part of their distributional areas.

The exact relationships within this complex will have to be clarified by hybridizations, particularly between the different diploids. Nevertheless, whatever may be the results of these experiments, *Claytonia* illustrates in a striking fashion the conservative influence of polyploidy. In spite of their very extensive chromosomal variation, these species are remarkably narrow in their overall range of morphological and ecological characteristics. All of their races are corm-forming inhabitants of deciduous woodlands in rich soil, which bloom in early spring and have a single pair of cauline leaves, a stereotyped architecture of their racemose inflorescences, and flowers, capsules and seeds which are very much alike. They represent multitudinous variations on a narrowly restricted theme.

The chronological age of mature polyploid complexes

Evidence from their distributional patterns indicates that most of the polyploid complexes which at present are in the mature state originated during the Pleistocene or Pliocene epochs, roughly between 500 000 and 10 000 000 years ago. In common with all evidence regarding problems of phylogeny, age and origin in groups without a fossil record, this evidence is indirect. Interpretations of modern distributional patterns are based upon the following assumptions, which are regarded as reasonable, in the light of ecological relationships between plants and their environment, and the migration of floras as indicated by the record of fossil pollens.

(1) Pleistocene and Pliocene distribution patterns were different from modern ones.

(2) During the Pleistocene epoch in North Temperate regions there were repeated oscillations of the flora southward and northward, in response to the oscillating climate.

(3) During the Pliocene epoch which preceded the glaciations, extensive mountain building activity in Eurasia, North America and South America disrupted previously continuous ranges, particularly of species adapted to mesic forest conditions.

(4) Since the development of initial and youthful polyploid complexes requires contacts and gene exchange between differently adapted diploids, the complexes which are now wide-spread and mature would have been able to acquire their present characteristics only if at an earlier time differently adapted diploids were sympatric and able to exchange genes with each other.

(5) New polyploids can extend their geographic distributions only if new habitats are opened up for them to colonize. Such events happened with particular frequency during the mountain building of the Pliocene and the retreat of glaciers during the Pleistocene.

Declining polyploid complexes

The decline of polyploid complexes first becomes evident through the existence of tetraploids or hexaploids which are not clearly related to any existing diploid. Although such situations might be explained on the assumption that the polyploids have diverged radically from their diploid ancestors by mutation and selection, the alternative explanation, that the diploid ancestor or ancestors have become extinct, is usually more compatible with other facts about the complexes concerned.

The decline of a polyploid complex is clearly evident in the Sections *Godetia* and *Biortis* of the genus *Clarkia*[139,165,181] (Figs. 6.9 and 6.10). Among the nine species in these sections there are three diploids with $n = 9$, one with $n = 8$, two tetraploids with $n = 17$, and three hexaploids with $n = 26$. None of the diploids resembles the two tetraploids closely enough to be regarded as ancestral to them, and only *G. speciosa* gives evidence of relationship to the ancestors of the hexaploids. The age of the complex is further evident from the fact that of the two tetraploids one, *C. Davyi*, is restricted to the coast of California, while the other, *C. tenella*, is a highly diverse collection of races or ecotypes found throughout temperate Chile and in western Argentina, being the only species of *Clarkia* in South America.

In this complex, the clearest evidence of extinction is the close relationship between the hexaploid *G. purpurea*, which is wide-spread and common in California, and the tetraploid *G. tenella*, which is equally wide-spread and common in Chile. This relationship is evident both from morphological resemblance and from chromosome pairing in their F_1 hybrid. The former existence in California of a tetraploid similar to *G. tenella*, which was one of the ancestors of *G. purpurea* is, therefore, highly probable. One of the diploid ancestors of *C. tenella* and its extinct Californian relative was probably similar to the modern *C. speciosa* ($n = 9$), but the other ancestor, which presumably had 8 pairs of chromosomes, is apparently not represented by any modern species. The 9-paired species which combined with

C. prostrata—26   C. purpurea ssp purpurea—26—C. purpurea ssp quadrivulnera   C. affinis —26

17

9

C. Davyi—17   C. tenella—17

9

9   8

C. Williamsonii—9

C. speciosa ssp speciosa—9   C. nitens—9

C. imbricata—8

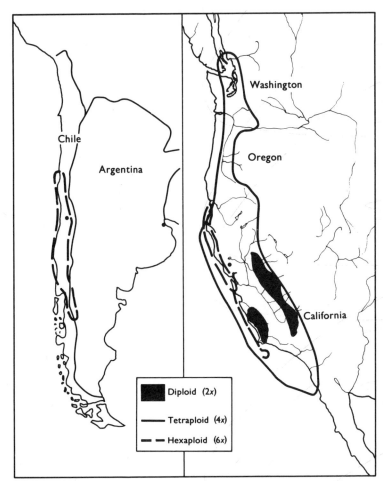

Fig. 6.10 (*above*)   Limits of distribution of diploids, tetraploids, and hexaploids of *Clarkia* sect. *Godetia.* (From Lewis and Lewis.[139])

Fig. 6.9 (*opposite*)   Representative plants and seed capsules of members of the polyploid complex of *Clarkia* subgenera *Godetia* and *Biortis.* The modern diploids all occur in California. The tetraploids ($n = 17$) are relatively rare in California but more common and much more variable in South America (Chile); the hexaploids are found throughout Pacific North America. At least one morphological characteristic, conspicuously pubescent capsules, occurs in the hexaploids and the South American *C. tenella,* but not in North American tetraploids or diploids. This suggests that a tetraploid or diploid possessing this characteristic, and ancestral to both *C. tenella* and the hexaploids, once existed in California but has become extinct. (From Lewis and Lewis.[139])

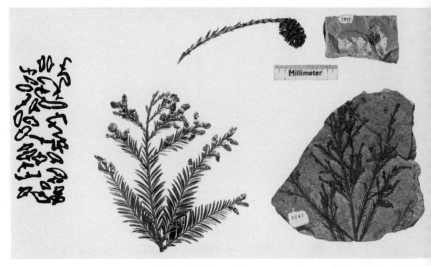

Fig. 6.11 Meiotic chromosomes (left); leafy branch with staminate cones and ovulate cone (middle) of modern *Sequoia sempervirens* (2n = 66); together with a leafy branch and an ovulate cone of the fossil *Sequoia* from the Oligocene Florissant Formation of Colorado. (Specimens from the collection of the Department of Paleontology, University of California, Berkeley).

the *tenella*-like tetraploid to give rise to *G. purpurea* is also probably extinct.

Relictual polyploids

The final stage of maturity of a polyploid complex is a high polyploid which has no close relatives. Several such examples are now known. One of them is the coast redwood of California, *Sequoia sempervirens*, which is a hexaploid (2n = 66). It is now regarded as a monotypic genus, since the other species previously assigned to *Sequoia*, '*S. gigantea*', is now recognized on the basis of many characters to belong to a different genus, *Sequoia-dendron*. Although it is diploid, *S. giganteum* is probably not ancestral to *S. sempervirens*. There are a number of fossil redwoods which look much more like *S. sempervirens* but which have relatively smaller leaves and cones, suggesting that they may have been diploids (Fig. 6.11).

Several other endemic species of California have chromosome numbers which indicate a polyploid origin, but are nevertheless monotypic genera or ditypic. Among them are *Lyonothamnus floribundus* (2n = 54), *Dendromecon rigidum* (2n = 56), and *Simmondsia chinensis* (2n = 52). A relictual polyploid, wide-spread throughout the northern hemisphere, is *Brasenia Schreberi*, of the Nymphaeaceae (2n = 80).

THE RELATIONSHIP BETWEEN POLYPLOIDY AND APOMIXIS

Several polyploid complexes are characterized by *apomixis*, which is defined broadly as the replacement of sexual by asexual reproduction. Sometimes this replacement is through the presence of vegetative buds or leafy rosettes in positions where flowers would normally be expected, but more often seed production is involved. In *Citrus, Opuntia*, and a few other genera, apomictic seed production is made possible by *adventitious embryony*, or the production of embryos by budding of the nucellus or other somatic tissue of the seed. More commonly, however, apomixis is a two-step process. Through various modifications of meiosis in megasporogenesis which give rise to megaspores having the unreduced chromosome number, or through the formation of an embryo sac by a somatic cell of the nucellus or ovular integument, gametophytes and egg cells arise having diploid, unreduced chromosome number. Development may then continue without fertilization by *parthenogenetic* development of the egg cell, or by *pseudogamy*, which includes fertilization of the polar nucleus to form a hybrid endosperm, but development of the embryo from the egg cell without fertilization. The details of these processes can be found in other books.[209]

In some groups, such as the genus *Citrus* and a few species of *Potentilla*, apomixis has developed in diploid species or hybrids. Much more commonly, however, it is associated with polyploidy. As with sexual polyploids, the genetic constitution of apomictic polyploids forms a complete spectrum from strict 'autopolyploidy' to 'allopolyploid' origin resulting from crossing between widely different diploid parents. Furthermore, when groups containing apomictic polyploids are studied in their entirety, the entire spectrum is usually found within the same circle of relationship. Good examples are the American species of *Crepis* (Fig. 6.12), the subgenus *Eubatus* of *Rubus*, and the genera *Poa* and *Bothriochloa* of the grass family.[27,53,208]

Secondary modification of apomictic complexes

Without full knowledge of the facts, one might expect that in polyploid complexes characterized by apomixis, secondary modifications would be less significant than in sexual complexes. This, however, is not the case, for several reasons.

Mutations of apomicts

In the first place, many apomicts are highly heterozygous genetically. This fact can be demonstrated by using the pollen of an apomict to fertilize a plant which is partly or entirely sexual. The resulting progeny

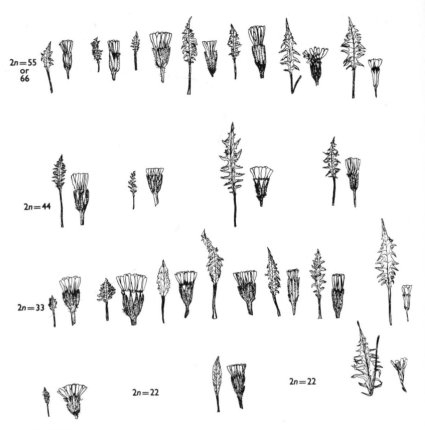

Fig. 6.12 Variability in leaves and capitula of a small sample of sexual species and apomicts of the *Crepis* apomictic complex in North America. Lower row, *C. modocensis* (left), *C. occidentalis* (centre) and *C. acuminata* (right): all diploid and sexual. Upper rows, apomicts showing various combinations of the characteristics of the three sexual species, and having somatic chromosome numbers of 33, 44, 55, and 66. (From Babcock and Stebbins.[7])

are always highly variable.[209] Because of this heterozygosity, many recessive as well as all dominant mutations alter the phenotype as soon as they appear. If, therefore, a single apomictic clone is propagated in large numbers for several generations, morphological variants always appear among its progeny.

Facultative apomixis

Secondly, many genotypes, known as **facultative apomicts**, are capable of either sexual or apomictic reproduction. For any particular clone grown in a constant, defined environment, the proportion of sexual to apomictic seeds is fairly constant. It can be determined most easily in genera like *Potentilla*[172] and some sections of *Poa*,[32] in which pseudogamous fertilization of the endosperm is required for development, and which therefore produce functional pollen. Presumably, whether a particular ovule will produce a haploid egg that can be fertilized or a diploid egg that develops apomictically depends upon a delicate balance of opposing factors which operate during development.

This balance can be modified by altering the external environment. In the grass species *Dichanthium aristatum*, for instance, continuous exposure to short days results in a high incidence of apomixis (87–93%) while exposure to only the minimum number of short days required for flowering causes a much lower percentage (55–63%) of apomixis.[124,125]

Geographic variations in the development of apomixis

In some groups, such as *Taraxacum*[77,78] and *Dichanthium*,[53] the dominant form of reproduction varies in different parts of the geographic range of the same species. The races of *Taraxacum officinale* found in northern and western Europe are nearly all obligate apomicts. Most of them are triploids having completely inviable pollen, in which the growth of the embryo begins even before the flowers open. In central Europe, however, many diploid sexual races of *T. officinale* exist, and hybridization between this species and other *Taraxaca*, such as *T. laevigatum*, has been recorded.[77] Many new apomictic biotypes appear in the progeny of such hybrids, some of which can spread outward from their centre of origin. Probably, therefore, the great diversity of races or microspecies of *Taraxacum* found in western and northern Europe, as well as through introduction into North America, has resulted from successive migrations outward from the central European centre. Direct observations have shown that even within this centre, the apomicts show more aggressive, weedy tendencies than the sexual biotypes.

Some genera which contain many apomicts do not exhibit such clear-cut patterns. In some species groups of *Poa* and *Potentilla*, for example, facultative apomicts exist in greater or lesser proportions almost throughout the range of the complex, and centres of concentration for diploid, sexual races have not been identified.

Comparisons between sexual and apomictic polyploid complexes

In view of the radical difference between, on the one hand, sexuality and cross fertilization and, on the other hand, asexual reproduction via apomixis, the similarity in both morphological variation pattern and eco-geographical distribution between sexual and apomictic polyploid complexes is striking. While young apomictic complexes have not yet been recognized as such, many mature complexes exist of which the diploids are relatively restricted in distribution and largely allopatric, while the polyploids are common, aggressive, and highly variable. The American species of *Crepis* (Fig. 6.13), and the genera *Taraxacum*, *Hieracium*, *Bothriochloa* and *Rubus* subg. *Eubatus* and *Antennaria* fit into this pattern. Furthermore, complexes like those of *Poa* and *Potentilla* resemble declining sexual complexes in that their diploid representatives are either extinct or hard to recognize as such, while higher levels of polyploidy in many areas predominate over tetraploidy. Finally one species *Houttuynia cordata*, which throughout its range is a high polyploid and is apomictic in all regions where it has been carefully examined,[7] represents a relictual polyploid apomict.

The principal differences between sexual and apomictic polyploid complexes lie in the chromosome numbers represented and in the local variation patterns. Since irregular meiosis is no barrier to their seed formation, triploid, pentaploid and other odd-numbered apomictic clones can be as successful in polyploid complexes as can tetraploids, hexaploids, and octoploids. Interestingly enough, aneuploid clones are not formed in most apomictic complexes. Exceptions are those of *Poa alpina* and *P. pratensis*, in which an extensive aneuploid series has long been known.[209]

The presence of asexual reproduction by apomixis sometimes brings about extreme genetic constancy. This, however, is not the usual situation if entire populations are considered. More often, they consist of collections of apomictic biotypes, which have slightly different microhabitat preferences, so that they do not compete directly with each other. Figure 6.14 shows the kind of pattern that one finds if such a population is scored for two distinctive morphological characteristics. One can conclude from this figure that the population consists of two dominant apomictic biotypes, and four or five rare ones, represented by only one or two individuals. The pattern is much like the one obtained when a population of predominantly self-fertilizing species is similarly examined.

Because of the constancy of many apomictic races or 'microspecies,' some botanists have given them taxonomic recognition. The difficulty with this procedure lies only partly in the fact that by it the number of described 'species' becomes much too large. In addition, the method leads to absurdity if it is extended to cover the entire extent of any apomictic complex. This is because at least some populations contain facultative apomicts, which are continually crossing with each other to yield new clones. If these clones are regarded as species, then different offspring of the same parent must occasionally be placed in separate species.

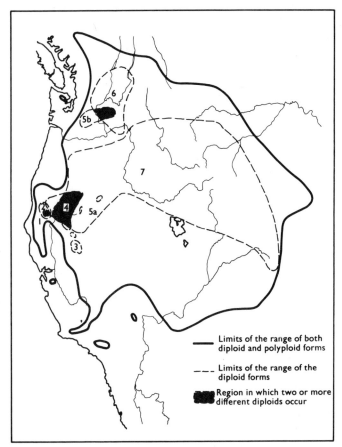

Fig. 6.13 Geographic distribution of the polyploid complex of *Crepis*. The three diploid sexual species illustrated in Figure 6.12 are nos. 3 (*C. occidentalis*), 5 (*C. modocensis*), and 7 (*C. acuminata*). (From Babcock and Stebbins.[7])

The evolutionary future of apomictic complexes

Apomictic complexes have often been characterized as evolutionary 'blind alleys.' The viability of this statement depends upon what meaning is intended by it. There is no reason for believing that the acquisition of apomictic reproduction will either shorten the evolutionary life of a group of species or restrict its ability to colonize a large number of habitats. This is because all known apomictic complexes contain some sexual or facultatively apomictic populations, which from time to time enrich the

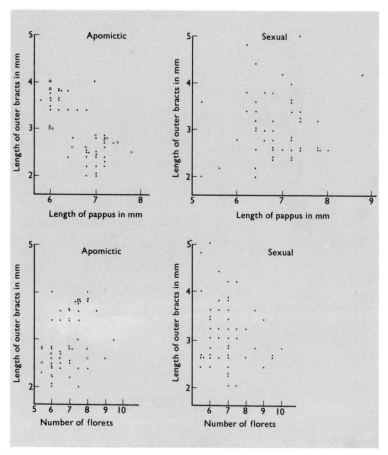

Fig. 6.14   Pattern of variability with respect to three morphological characteristics in an apomictic population of *Crepis acuminata* (left), as contrasted with a sexual population (right). In the upper scatter diagrams, length of outer involucral bracts (phyllaries) is compared with length of pappus; in the lower diagrams it is compared with number of florets. Note that the total range of variability in the apomictic population is less than that in the sexual population, but that the individuals of the former are grouped into two distinct clusters of variants. Individuals which on the basis of a combination of characteristics did not appear to belong to either of the two apomictic clones represented by the clusters are indicated by open circles. (From Babcock and Stebbins.[7])

gene pool represented by the obligate apomicts.[31] Distributional evidence suggests that some apomictic complexes are millions of years old. This is almost certainly true of the blackberries, *Rubus* subg. *Eubatus*, among which exist groups of Old World apomicts that are related to those of the New World. This complex must have arisen and spread along with the Arcto-Tertiary flora during the Tertiary period. The apomictic complexes of tropical genera like *Panicum*, *Setaria*, and *Pennisetum* are likewise most probably of Tertiary origin, since tropical floras were probably not affected enough by Pleistocene changes in climate to afford opportunities for new polyploid apomicts to become established and spread. These complexes are now so wide-spread and successful in a variety of habitats that their extinction in the foreseeable future is hard to imagine.

On the other hand, the hypothesis that apomictic complexes will not give rise to new genera or families, or even to characteristics not present in their sexual diploid ancestors, is very plausible. The principal support for this hypothesis lies in the fact that all known apomictic complexes are contained within the same genetic comparium, in that their range of variation is bounded by a group of sexual or facultatively apomictic species which are capable of hybridizing and exchanging genes with each other. In certain large genera, such as *Poa* and *Hieracium*, the apomicts belonging to different subgenera have completely different mechanisms for circumventing meiosis. This suggests that in these genera apomixis has arisen more than once, and that their apomicts have not been able to extend their patterns of variation beyond the confines of a single section or subgenus. Most polyploid complexes, both sexual and apomictic, generate chiefly an endless number of variations upon a single adaptive theme. Nevertheless, some sexual complexes, through genetic diploidization and diversification of certain polyploids, have been able to break through the boundaries originally established by their diploid ancestors, and have given rise to new genera or even families. Once apomixis has set in, even this possibility is denied to the evolutionary line. It is doomed forever to generate one variant after another on the old, original theme. It will survive only as long as some of these variants are still adaptive in some habitat.

POLYPLOIDY AND PLANT DISTRIBUTION

There are three reasons why analyses of polyploid complexes can provide particularly valuable evidence for interpreting patterns of distribution in terms of past migrations of floras and origins of species. In the first place, the phylogenetic progression from lower to higher numbers, while not completely irreversible, is probably more so than is any other phylogenetic series which can be studied on the basis of modern forms. More important, in the case of young complexes as well as most mature ones, the phylogenetic progression can be duplicated in the laboratory and garden by artificial hybridization and chromosome doubling. Finally, in many mature complexes the diploid representatives are now widely separated from each other, but polyploids exist which can be demonstrated experimentally to contain combinations of their chromosomes and genes. This is strong evidence in favour of the hypothesis that such diploids once occurred sympatrically with each other, so that they were able to hybridize and exchange genes. By putting this evidence together with that derived from fossil floras which indicate the past distribution of certain kinds of plant communities, a synthesis can often be achieved which sheds much light upon past migrations and alterations of floras. In order to do this, we must consider the relationship of polyploids to climatic, edaphic, and historical factors.

Polyploidy and climate

Early studies of the geographic significance of polyploidy emphasized its relationship to major climatic differences. The hypothesis was advanced that polyploids are better adapted than their diploid progenitors to extreme cold and drought. The difficulties with this hypothesis have already been pointed out by several authors.[209] The data regarding cold resistance were obtained chiefly from Europe, where percentages of polyploidy in the flora as a whole definitely increase as one goes northward to higher latitudes. On the other hand, the percentage of polyploids in the flora of the high Alps, comprising the rocky refuges for plants which exist above the snow line, is no higher than in the plains and foothills below.[69]

The hypothesis that desert regions contain high percentages of polyploids was based upon a few examples taken from the flora of Timbuktu. More recent studies of the floras of arid zones fail to confirm this hypothesis. Particularly enlightening are comparisons of the distribution of diploid and polyploid populations belonging to groups which are distributed both in desert regions and in the more mesic regions adjacent to them. Such comparisons, which are summarized for western North America in Table 6.2, give no support to the hypothesis that polyploids are in general more tolerant of drought than their diploid ancestors. Furthermore, physiological comparisons of artificial autopolyploids with their diploid progenitors have shown that such polyploids have a decreased resistance to drought.[209]

The hypothesis that equable climates favour diploids and extreme climates favour polyploids has been finally made untenable by the results of cytological investigations of tropical rain forest floras, particularly that of the Ivory Coast,[154] the fern flora of Ceylon,[157] and certain families (Labiatae, Commelinaceae) in West Africa[166,167] and India.[180] If genera having high basic numbers are regarded as originally polyploid, the

Table 6.2 Comparison between pairs of diploid and closely related polyploid races or species, one or both of which are found in desert or steppe regions of the western United States. In only one out of nine examples is the tetraploid found in more xeric habitats than its diploid relatives.

Genus and Reference	Diploid	Ecological occurrence	Polyploid	Ecological occurrence	Comparison polypl/2x
Ambrosia, Payne et al.[178]	A. deltoidea	Deserts	A. chenopodiifolia	Dry hills, coastal	More mesic
Artemisia, Ward[237]	A. tridentata, cana, arbuscula	Dry open steppes	Same, plus A. Rothrockii	Dry open steppes or dry forests	Similar or more mesic
Castilleja, Heckard[99]	C. chromosa	Dry open steppes	Same, plus C. affinis	Steppes, chaparral, forests	Similar or more mesic
Dudleya, Uhl and Moran[234]	D. Abramsii, D. saxosa aloides	Deserts, desert margins	D. lanceolata, D. saxosa saxosa	Desert margins, coastal chaparral	Similar or more mesic
Eriogonum, Stebbins[208]	E. fasciculatum ssp. polifolium	Deserts, desert margins	E. fasciculatum ssp. foliolosum	Chaparral, interior and coastal	More mesic
Galium, Dempster and Ehrendorfer[50]	G. Mathewsii, magnifolium, multiflorum, coloradoense	Desert margins, dry forests	G. argense, Munzii, hilendae	Desert margins, dry forests	Similar
Gutierrezia, Solbrig[201]	G. microcephala	Deserts	G. bracteata, californica	Desert margins, chaparral	More mesic
Haplopappus, Raven et al.[182]	H. acradenius ssp. bracteosus	Semi-arid valleys	H. a. acradenius	Deserts	More xeric
Sidalcea, Hitchcock and Kruckeberg[104]	S. multifida	Open steppes	S. glaucescens	Dry forests	More mesic

percentage of polyploidy in these floras ranges from 85 to 95 per cent. Present data suggest that the lowest percentages of polyploidy are found in floras of warm temperate and subtropical regions, and that percentages increase as we go from these regions either toward the cooler or the tropical areas.

Edaphic and historical factors affecting the distribution of polyploids

More significant than the correlations with climate are correlations with degree of disturbance as well as with the age of habitats. In compiling the data for such correlations, one cannot deal with floras as wholes, because of the great differences in the frequency of polyploidy in plants having different life cycles and growth habits and particularly because the proportion of plants having different life cycles and growth habits differs greatly from one habitat to another. In particular, many habitats which have been recently opened up by human activity, such as old fields and roadsides, contain unusually high proportions of annual species, among which polyploids are much less frequent than among perennial herbs (Table 5.1). Consequently, the fact that in Canada the percentage of polyploids among weeds is the same as in the flora as a whole[170] does not tell us anything about the relative probability of becoming weedy among related diploids and polyploids which have the same growth habit. Significant facts about the distribution of polyploids are much more likely to be obtained from comparisons within groups of related species and chromosomal races than from statistical comparisons of entire floras. The results of representative intra-group comparisons will be given for plants having various habits of growth and occupying areas which have been subjected to various kinds of disturbances.

Annual weeds of fields and roadsides

The most reliable data on these plants can be obtained in the case of species groups which are native to regions recently opened to cultivation, and have therefore acquired their weediness in modern times. Seventeen such groups exist in the California flora, and among them there are 36 species which have become weedy. Of these 15, or 42 per cent, are polyploid, which compares with a figure of 31 per cent polyploidy for the flora of California as a whole, and 17 to 20 per cent for species of annuals with known chromosome numbers. Thirteen of these species have diploid non-weedy relatives in California, one (*Bromus arizonicus*, see p. 133) is a high polyploid having ancestors with lower degrees of polyploidy elsewhere in America, and the remaining tetraploid species (*Calandrinia ciliata* var. *Menziesii*) has Californian relatives of which chromosome numbers are not known. A nearly complete list of these species is given elsewhere.[215] None of the diploid annual Californian weeds has polyploid relatives which are not weedy.

These data lead to the conclusion that diploid annuals belonging to genera not containing polyploid annuals have a good chance of becoming weedy, but that if related diploids and polyploids exist in the same group of annuals, the polyploids have a greater chance of becoming wide-spread as weeds than their diploid relatives. This conclusion is supported by the distribution pattern of the genus *Aegilops*, which has already been described (Fig. 6.3).

Polyploidy and edaphic factors

The bearing of edaphic factors on the relative distribution of diploids and polyploids has received far less attention than it deserves. The immediate availability of habitats for colonization by newly formed polyploids depends much more upon such factors as the nature of the soil, local differences in temperature, and the density and character of existing vegetation than it does upon the overall climate. Hence we need to know much more than we do about the relative frequency of polyploids in different habitats within the same climatic zone.

The promising results which can emerge from such studies are well illustrated by Johnson and Packer's analysis[114] of the Ogotoruk Creek Flora of northwestern Alaska. In this arctic flora, from 68° north latitude, an environment gradient scale was established on the basis of soil characteristics. When the frequency of polyploids was compared between plant communities occupying different positions on this scale, the highest frequency (88%) was found in lowlands having fine textured soils with high moisture, low temperature, shallow permafrost and high disturbance (Fig. 6.15). At the other end of the scale the communities occupying soil having coarse texture, low soil moisture, high soil temperature, deep permafrost and low disturbance had only half as many polyploids (43%). Since in this arctic climate annuals as well as trees are virtually absent, the communities examined were sufficiently similar in respect to the life forms represented that variations due to this factor could not have affected the results.

Similar data need to be obtained on floras of other regions. Even more significant data could be obtained by confining the observations to those genera having different levels of ploidy in the same geographic region.

Polyploidy and glaciation

The observations of Johnson and Packer were made in order to provide information regarding the effects on the frequency of polyploidy in the

North Temperate Zone of the Pleistocene glaciations themselves as well as of associated disturbances which occurred in surrounding unglaciated regions. Although the Ogotoruk Creek area was itself not glaciated, its habitats certainly suffered great changes during the Pleistocene, the magnitude of the changes having been greatest with respect to the communities at the left hand end of the scale in Figure 6.15. The results of this study, therefore, support the conclusions that polyploids are at a selective advantage in habitats which have been subjected to frequent and drastic fluctuations in both climate and edaphic factors.

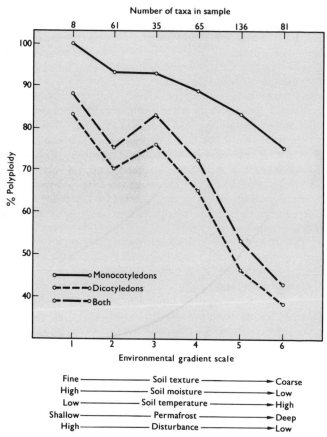

Fig. 6.15 The relationship of the frequency of polyploids in the angiosperm flora of the Ogotoruk Creek Valley, Alaska, to environmental gradients of soil character, moisture, and persistence of frost. (From Johnson and Packer.[114])

The mistaken notion that polyploidy has evolved in response to increasingly cold climates arose because the first comparative data made available were those obtained from western Europe and the islands to the north of this continent. In these regions, three different factors vary with increasing latitude to the same degree and in the same places. These are increasing cold, increasing effects of glaciation (except possibly for a few relatively small nunatak areas), and increasing isolation due to insularity. Along the Pacific Coast of North America, these factors are separated from each other. Insularity need not be considered, since such islands as exist are so close to the continent that seed dispersal to them presents no problem. The severity of climate along this coast increases most drastically between 53° and 68° north latitude. At 53° (Queen Charlotte Islands), the mean temperature for the coldest month is about 0°C, and the abundant moisture supports a luxuriant forest, at lower altitudes. At 68° (north-western Alaska), winter temperatures reach −40°C, and permafrost extends to depths of over 300 m. On the other hand, the effects of glaciation are already strong at 53°, and are little, if any, stronger at 68°, since this region of north-western Alaska was largely unglaciated. The major difference between regions little affected and those much altered by the glaciers themselves as well as by their influence upon surrounding territories occurs between 40° and 50° north latitude.

Estimated percentages of polyploidy in the Pacific Coast flora of North America are 36 per cent for northern coastal California at about 40°, 53 per cent for the Queen Charlotte Islands at 53°[229] and 56 to 59 per cent for north-western Alaska at 68°.[114] These data show that the maximum increase in percentage of polyploidy is associated not with the maximal change in climate, but with the greatest difference with respect to the effects of glaciation.

Further information on the relationship of polyploidy to glaciation can be obtained by comparing the distribution of diploids and polyploids belonging to individual complexes with the extent of glaciation in the regions where they occur. That of *Biscutella laevigata* and its relatives in the Alps and the surrounding mountains is particularly instructive (Fig. 6.16). Diploid races or subspecies have persisted in montane refuges both to the west and the east of the major glaciated area of the central Alps, the latter being occupied almost entirely by tetraploid populations. These surrounding unglaciated or little glaciated mountains are refuges for diploids in many other groups which have contributed polyploids to the flora of the glaciated area.[65,70] In North America, the distribution of *Iris versicolor* ($2n = 108$) and its probable diploid ancestors, *I. virginica* var. *Shrevei* and *I. setosa* var. *interior* (Fig. 6.17), is equally instructive.

To produce situations like these, the following sequence of events probably took place. Before any particular advance of an ice sheet, the worsening climate opened up habitats for the hardier members of the

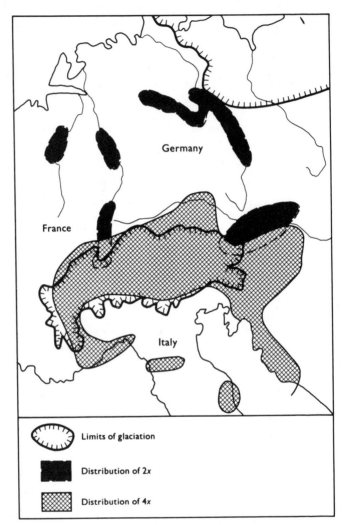

Fig. 6.16 The association between the distribution of diploid and tetraploid *Biscutella laevigata* and the Pleistocene glaciation of Central and Northern Europe. The diploids occur chiefly in unglaciated river valleys between regions covered by the northern and the alpine glaciers. The tetraploids occur throughout the area covered by the alpine ice sheet, and extend from there southward and eastward. (From Manton.[155])

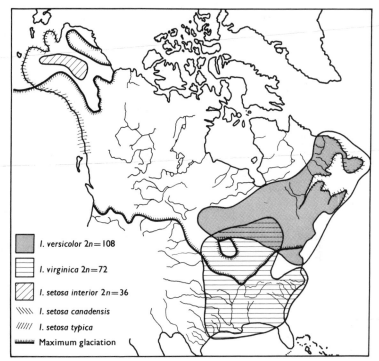

Fig. 6.17 The distribution of the hybrid polyploid *Iris versicolor* and its probable ancestors in relation to the Pleistocene glaciation of North America. The distribution of *I. versicolor* is almost entirely within glaciated territory; *I. virginica* is in both glaciated and unglaciated regions; *I. setosa* var. *interior* occurs in the unglaciated regions of central Alaska. Its former occurrence along the southern margin of the ice sheet in eastern North America is postulated. (Data of E. Anderson, from Stebbins.[209])

Legend within figure:

- *I. versicolor* 2n=108
- *I. virginica* 2n=72
- *I. setosa interior* 2n=36
- *I. setosa canadensis*
- *I. setosa typica*
- Maximum glaciation

group, through destruction of the less hardy members of the pre-existing vegetation. In a mountainous region, this destruction would have varied greatly according to slope, insolation, and the character of the soil. Consequently, differently adapted populations would have in many places become established near each other, and hybridization between them would have been possible. With the formation and advance of the glaciers themselves, these variably distributed conditions would have become general around the margins of the ice sheet, and would have persisted throughout the glacial period. Consequently, many hybrids as well as polyploids having various degrees of hybrid origin probably existed around these margins when the retreat of the ice began. Geological evidence indicates that in both montane and flat areas this retreat was highly irregular, so that in many places adjacent areas would have supported on the better and warmer soils the beginnings of forest associations, and on the poorer, colder soils the remnants of the retreating tundra vegetation. These conditions would have provided more opportunities for hybridization between diploids, between different ones of the newly formed polyploids, and between polyploids and their diploid ancestors. From these rich, variable gene pools the genotypes best adapted to the habitats being opened up by the retreating ice would have become established and spread by natural selection. Most of these were polyploids, but in some groups diploids also became adapted to these new conditions.

Effects on polyploidy of pre-glacial environmental changes

By far the majority of the polyploids of which the ancestry can be determined with any degree of certainty have distribution patterns which suggest an origin during the Pleistocene or recent epochs, or perhaps during the end of the Pliocene epoch, when the climate was deteriorating prior to the advance of the ice sheet. This is true of virtually all of the analysable polyploid complexes of California, which have been listed elsewhere[220] as containing 'patroendemics' or 'apoendemics'. Nevertheless, many polyploids have distributional ranges which cannot easily be explained on this basis. This is particularly true of polyploids which are native to temperate or warm regions of both the Old and New World, as well as of those which are confined to one continent, but have at least one ancestor now existing only on another continent. Several polyploid species of ferns are known from both Eurasia and North America, of which *Asplenium trichomanes* is a good example. Among seed plants, the hexaploid race of the grass *Heteropogon contortus*, found in subtropical Africa, Asia and North America, and tetraploid *Themeda quadrivalvis*, found in Africa, Asia and Australia, are both good examples. These species must have acquired their distributional ranges at a time when migrations from one continent to another were accomplished more easily than they are at present.

The same is true of examples such as the species of *Lactuca* and *Crepis* endemic to North America.[7,42] Those of *Lactuca* all have the gametic number $x = 17$, and have relatives in Asia with $x = 8$ and $x = 9$. The American species of *Crepis*, which form a polyploid series based upon $x = 11$, probably originated as hybrid polyploids between different Asiatic species having $x = 4$ and $x = 7$. Throughout the Pleistocene and recent epochs the climates of the Siberian–Alaskan land bridge which connects Asia with America have been arctic or subarctic, and too inhospitable for these temperate plants. However, evidence from fossil pollen indicates that this land bridge could have been suitable for them during the middle of the Tertiary period, when these polyploids probably arose. This hypothesis is compatible with the amount of diversification found in these two species groups, which can then be explained in relation to changes in the flora which took place during the latter part of the Pliocene as well as the Pleistocene epoch.

The principal kinds of environmental changes which took place during most of the Tertiary period and in some parts of the world during the Cretaceous period, and could have provided conditions favourable for the origin and spread of polyploids, were the following. Extensive mountain building brought into existence many new alpine and subalpine habitats.[5] At the same time, this activity produced many 'rain shadow' areas in the interior parts of the continents, which consequently became much drier, and developed continental climates having extreme seasonal differences in temperature. In addition, the Cretaceous and Tertiary periods witnessed a succession of evolutionary appearances of new kinds of grazing animals. Grazing dinosaurs, large flightless birds (Ratitae), primitive ungulates, Titanotheres, elephants, horses, camels, antelopes, sheep and bovids all arose at various times. Both by their different methods of grazing and browsing, and the different degrees to which they trampled the soil, they must have created new and sometimes drastic pressures upon the existing vegetation. Moreover, these animals must have often transported seeds from one region to another in their hair or in the mud adhering to their legs and feet. The catastrophic effects which are exerted on vegetation by the introduction of new grazing or browsing animals that carry with them the seeds of new kinds of plants are evident both from the almost complete replacement of native herbs by introductions from Eurasia during historical times in America and Australia, and from the damage which the forests of Hawaii and other Pacific islands have suffered, and those of New Zealand are still suffering, from similar invasions. Although the association of individual appearances of polyploidy in particular groups with specific examples of these changes may never become possible, we must be aware of the fact that they have been taking place in various places from time to time throughout the evolutionary history of the flowering plants. We can thus expect to find polyploid complexes of various ages, a conclusion which has already been reached on the basis of comparisons between the distributional patterns of individual complexes.

Polyploidy and endemism

A subject which has always been of great interest to plant geographers is the occurrence of endemic species restricted to narrowly confined areas of distribution. Although in the past great controversies have arisen between some botanists who thought that all endemics are of recent origin and others who believed that they are ancient relics, most modern plant geographers recognize that some endemics may be recent, others of moderate age, and still others may be ancient or relictual. A combination of palaeobotanical and floristic evidence supports this latter point of view. This evidence indicates that narrowly endemic species of trees in temperate regions are ancient relics. Many endemics of arid or high montane regions must be of moderate age, since habitats were opened up for them during the middle or latter part of the Tertiary period. Still other endemics, which are localized in specialized habitats in regions like western North America and the Mediterranean region, and have close relatives of more generalized distribution in the surrounding vegetation, must be of recent origin.

Polyploids contribute to this problem because of the usually irreversible trend from lower to higher levels of polyploidy. Based upon this consideration, two kinds of endemics have been recognized in genera having different levels of polyploidy.[71] *Patroendemics* are narrowly restricted diploids having polyploid relatives with wider distributional areas. *Apoendemics* are narrowly restricted polyploids, usually at levels higher than tetraploidy, which are related to and probably descended from wide-spread diploids or species having lower levels of polyploidy. The frequency distribution in different parts of California of patro- and apoendemics belonging to 52 different polyploid complexes is shown in Figure 6.18. Patroendemics are seen to be concentrated chiefly in the coastal zone. The maritime climate found in this zone, with mild winters and cool summers, was much more wide-spread in western North America during the Tertiary period, judging from palaeobotanical evidence.[5] Moreover, many of the woody species which are now largely or entirely confined to that zone formerly extended much further inland. The apoendemics are more evenly distributed. Their high number in the central coast reflects both the great diversity of habitats found in that area and the abundance of diploid species from which they could have arisen. In the inner coast ranges, the Sierra Nevada, and the desert mountain ranges apoendemics considerably outnumber patroendemics. These regions have suffered great changes in climate, particularly toward varying degrees of aridity and continentality. In addition, some apoendemics have occupied areas made available by the retreat of the glaciers which formerly occupied many parts of the Sierra Nevada.

SECONDARY CYCLES OF POLYPLOIDY

A characteristic feature of old, mature, or declining polyploid complexes is the ability of certain segments of a complex to initiate new series of polyploid numbers, in which the basic number that is multiplied is not the original basic number of the complex, but some multiple of it. These

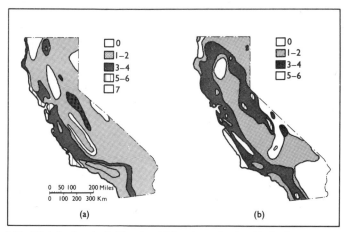

Fig. 6.18 Distribution in California of (**a**) patroendemics (diploid representatives of polyploid complexes) and (**b**) apoendemics (high polyploids related to diploids or lower polyploids living in the same region). Note that the highest number of patroendemics is in the central coast, which has always had a relatively mild climate but where climatic oscillations during the Pleistocene epoch promoted the formation of new habitats into which polyploids derived from these diploids could move. The regions containing higher numbers of apoendemics are more widely scattered, but are conspicuous in the inner Coast Ranges and Sierra Nevada. These regions have in more recent geological epochs acquired relatively harsh climates, with hot dry summers and colder winters; and in the Sierra Nevada apoendemics are characteristic of glaciated areas. (From Stebbins and Major.[220])

new series can be called secondary cycles of polyploidy. Some of these cycles begin at a time when species having the original basic number of the complex still exist, though usually in a different area and most often on a different continent. In that case the new cycle merely proliferates species belonging to the original genus. In other instances, they begin at a declining or relictual stage of the original complex. This produces a group of closely related species which show a marked morphological discontinuity from any others of their relatives. From the taxonomic point of view they are then recognizable as a distinct genus. The basic chromosome number of this genus, that from which the new cycle of polyploidy has started, is consequently of secondary polyploid origin. A few examples, selected from a large series of possible ones, illustrate how this transition might come about.

Secondary cycles within a genus

A good example of secondary cycles of polyploidy within a genus is in

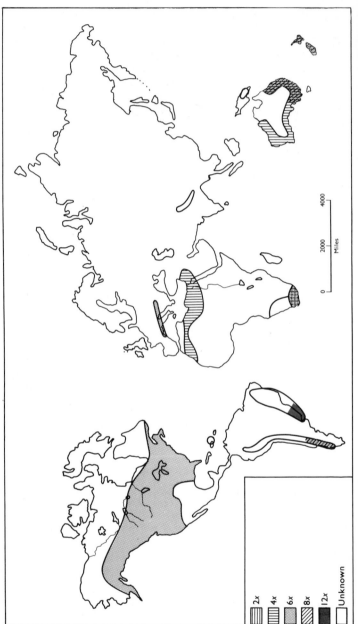

Fig. 6.19 Distribution of chromosome numbers in the genus *Danthonia*. The diploid species, now confined to South Africa, are probably relictual. The hexaploid species of Europe, Asia (? not yet counted), North America and South America belong to the same section, and are probably the disjunct remnants of a former holarctic distribution. Those of Australia – New Zealand and South America are not closely related to each other, so that an antarctic dispersal of the genus is questionable.

2x
4x
6x
8x
12x
Unknown

0   2000   4000

Miles

Danthonia, of the grass family (Fig. 6.19). The original basic number of this genus is $x = 6$, as is evident from the presence in South Africa of several diploid species having the somatic number $2n = 12$.[51] By far the commonest somatic numbers in the genus are, however, $2n = 24$ (tetraploid) and $2n = 36$ (hexaploid). These polyploid levels are found on different continents. In Australia, where the genus is dominant and contains a large number of species, the principal numbers present are $2n = 24$ and $2n = 48$. These tetraploid and octoploid species form a typical polyploid complex, of which the basic gametic number is $x = 12$. The few Australian species having $2n = 36$ belong to a different section of the genus, and apparently have not participated in the polyploid complex.

In North and South America, on the other hand, the only known chromosome numbers are $2n = 36$ and $2n = 72$. Since the only European species, *Danthonia calycina*, has $2n = 36$ and resembles closely some of the North American species, the hexaploids probably represent the result of a holarctic dispersal which took place during the Tertiary period, and extended to South America. The South American species having $2n = 72$ are probably derived from North and South American hexaploids.

Genera with secondary basic numbers of polyploid derivation

Some genera have basic numbers which are obviously of secondary polyploid derivation. They are well represented by *Epilobium*, *Chamaenerion*, and *Zauschneria* of the Onagraceae (Fig. 6.20). Nearly all of the species of the world-wide genus *Epilobium* have the gametic number $n = 18$ ($2n = 36$), which in most continents appears to be its basic number. There are, however, two annual species in western North America, *E. paniculatum*, which has $2n = 24$, and *E. minutum*, which has one subspecies with $2n = 26$ and another with $2n = 32$. This suggests that the original basic

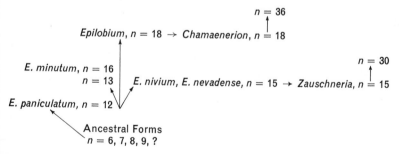

Fig. 6.20 Diagram showing the probable phylogeny of chromosome numbers in the genus *Epilobium* and its relatives in the family Onagraceae. Further explanation in the text. (From information provided by Peter Raven.)

number of *Epilobium* was either $x = 6$ or $x = 9$, and that $x = 18$ is of secondary polyploid origin. In *Chamaenerion*, which many systematists regard as a subgenus of *Epilobium*, there is a polyploid series of entities having the gametic numbers $n = 18$, $n = 36$ and $n = 54$, which includes forms described as tetraploid 'races' of both *C. angustifolium* and *C. latifolium*[169,196] plus hybrids between them.[18] This can be regarded as a youthful polyploid complex based upon $x = 18$.

A different basic number, $x = 15$, is found in *Zauschneria* and in three narrowly endemic species of *Epilobium*, *E. nivium* of northern California, *E. nevadense* of southern Nevada, and *E. suffruticosum* of Idaho and Wyoming. In their vegetative chracteristics, all of the species with $n = 15$ resemble each other so closely that they most probably had a common origin. *Zauschneria* differs from the three species of *Epilobium* mentioned above chiefly in its scarlet flowers having long calyx tubes, an obvious specialization for pollination by hummingbirds. Although the relationship between the numbers $n = 18$ and $n = 15$ is unclear, both of them are most probably of secondary polyploid origin. Hence the polyploid complex of *Zauschneria* (Fig. 6.4) probably represents a secondary cycle of polyploidy.

The genera *Epilobium*, *Chamaenerion* and *Zauschneria* are so closely interrelated that conservative botanists might be inclined to place them all in the same genus. Similar cytological relationships exist, however, between genera which are more distinct from each other. A very good example is the pair of genera *Primula* and *Dodecatheon*. Most of the species belonging to the large Eurasian and American genus *Primula* have the gametic number $n = 11$, but a few tetraploids exist with $n = 22$, while the group of *P. farinosa* forms an extensive polyploid complex based upon $x = 9$. The genus *Dodecatheon* is very similar to *Primula* in growth habit, differing chiefly in the shape of its flowers. Its species form a polyploid series based upon $x = 22$, with somatic numbers of $2n = 44$, $2n = 88$, and $2n = 132$.[230] They are confined to North America, where they occupy many of the ecological habitats which in Eurasia support species of *Primula*. The origin of *Dodecatheon* was most probably due to the extensive diversification of a few tetraploid derivatives of *Primula* followed by a second cycle of polyploidy.

Chromosome numbers of woody angiosperms

The series of examples presented above illustrate successive stages in the evolution of distinct genera having secondary basic numbers and cycles of polyploidy. In addition, they form an instructive background for interpreting the differences in chromosome number between woody and herbaceous angiosperms. In the floras of the temperate zone, trees and shrubs have higher basic numbers and, on the average, lower frequencies of polyploidy within a genus than perennial herbs (Table 5.1). Although tropical floras are still much more poorly known cytologically than temperate ones, such data as are available indicate that this difference does not hold for them. Investigations of herbaceous families with primarily tropical distributions, such as Acanthaceae,[87] Commelinaceae,[167] Bromeliaceae[158] and Orchidaceae,[117,118] reveals a series of basic numbers just as high as those found in temperate and most tropical groups of woody plants, and a tendency for subgenera, genera, tribes or even entire families to be uniform in chromosome number. When differences in ploidy level exist within the same tropical group, the species exhibiting them are usually easily distinguishable on morphological grounds. Chromosomal races within taxonomic species are rare in the tropics, and are confined almost entirely to ruderal species and weeds about human habitations. On the other hand, woody plants of tropical regions resemble those of the temperate zone in the rarity of polyploid series within a genus, and their basic numbers are similar, with a mode at $x = 11, 12, 13$, and 14. Nevertheless, among the tropical woody families which have been investigated, two, Annonaceae and Dipterocarpaceae[119] have basic numbers of $x = 7$, 8, and 9, which are the mode for temperate herbs.

These facts suggest that the factors responsible for the differences in chromosome numbers between woody and herbaceous genera found in temperate floras are not developmental and physiological but ecological and historical. Darlington[41] and others formerly advanced the hypothesis that polyploids are uncommon in woody plants because the increase in cell size associated with polyploidy decreases the efficiency of their cambial tissues, while the present author[207] once suggested that this increase would disturb the size relationships of cells to such an extent that efficient wood fibres could not become differentiated. Mangenot and Mangenot,[154] however, have pointed out that the highest modal range of chromosome numbers $(2n = 56$ to $2n = $ c. $144)$ known in any angiosperm family exists in the Bombacaceae. This family consists entirely of large trees, which are highly efficient in their adaptation to both wet rain forests and dry savannas. Moreover, high chromosome numbers exist even in temperate woody species, such as $2n = 152$ in *Salix* (3 species), $2n = 84$ in *Betula lutea*, and $2n = 104$ in *Acer rubrum*.

An alternative explanation for the low frequency of polyploid series in temperate as well as tropical woody genera is more consistent with the ecological relationships of polyploids described earlier. Throughout the Tertiary period, and in the case of tropical woody genera for perhaps even a longer period of time, these groups have not had to cope with drastic environmental differences. As components of climax stages of plant succession, they have advanced into new regions only when both climatic and soil conditions have become similar to those in their previous homes. Consequently, if occasional polyploid individuals have arisen, they have not found any habitat in which they had an adaptive superiority over their diploid progenitors. If drastic environmental changes followed by the opening up of new habitats is the principal stimulus for the establishment and success of polyploids, as is indicated by data already presented in this chapter, the low frequency of polyploid series in woody genera can best be explained on the basis of the fact, which is well documented by palaeo-botanical evidence,[28] that ever since the middle of the Tertiary period woody plants have been decreasing in numbers and abundance, while herbaceous groups have been correspondingly expanding.

This hypothesis is supported by the ecological position of the exceptional woody genera which do have a high percentage of polyploids. Throughout the glaciated regions of the northern hemisphere, the woody angiosperms that have been most conspicuously successful as invaders of newly available habitats have been the willows (*Salix*) and birches (*Betula*). These genera both have high percentages of polyploidy (50% for *Salix* and 42% for *Betula*), and by far the greatest concentration of polyploid species and races is in the glaciated regions. Woody angiosperms that have been particularly successful in colonizing open habitats created by human destruction of forests have been the dog roses (*Rosa* sect. *canina*) in Europe, the hawthorns (*Crataegus*) in North America, and the brambles (*Rubus*) in both hemispheres. All of these genera have very high percentages of polyploidy, reinforced in *Crataegus* and *Rubus* by apomixis.

The high basic numbers of woody plants and tropical herbs

A paradoxical fact is that in temperate floras woody genera, in spite of their lower incidence of polyploid series, nevertheless have significantly higher basic numbers than herbaceous genera (Table 6.3). Based upon the well known fact that woody plants are less specialized anatomically than herbs, two explanations can be offered for this difference. One is that the original basic number of angiosperms was in the modal range for modern woody genera, i.e., $x = 12$, 13, or 14, and that the basic numbers of herbs were derived by descending series of aneuploidy, such as were described in Chapter 4. The other explanation is that basic numbers of modern woody genera were derived by ancient polyploidy, and that the original basic numbers of angiosperms, both woody and herbaceous, were $x = 6$ and $x = 7$.

The evidence now available strongly favours the second hypothesis. This evidence is of two kinds. In the first place, most woody families of angiosperms are now known to include a few genera having basic numbers of $x = 7$, $x = 8$, and $x = 9$. The great majority of these genera are tropical, in agreement with the widely accepted hypothesis that temperate

Table 6.3 Comparison between the basic gametic numbers of angiosperm genera, classified according to predominant growth habit and temperate or tropical distribution. Those containing high proportions of both herbaceous and woody species were omitted from the list. They were classified as temperate or tropical on the basis of whether the majority of species are distributed within the tropics or outside of them. Those having high proportions both of temperate and of tropical species were omitted. Basic chromosome numbers were taken as the lowest known number in the case of genera having multiples of a single basic number. In the case of aneuploid series, the modal number which forms the basis of polyploid series was scored. Genera having extensive aneuploid series, such as *Carex* and *Stipa*, were not included. The quartiles were made as nearly equal in number of genera as possible, but that of the highest range was made somewhat smaller than the others, because of the wide range of basic numbers included in it. Additional cytological information on many tropical genera may shift the numbers for them toward higher frequencies for the lower basic number, but radical alteration of the nature of this chart through such information is unlikely. (Data chiefly from Darlington and Wylie.[42])

	$x = 3$ to $x = 8$		$x = 9$ to $x = 11$		$x = 12$ to $x = 16$		$x = 17$ or higher		
	No. of Genera	Percentage	No. of Genera	Percentage	No. of Genera	Percentage	No. of Genera	Percentage	Total
Temperate, herbaceous	472	38	347	29	215	18	182	15	1216
Tropical, herbaceous	53	15	121	34	94	26	92	25	360
Temperate, woody	32	9	91	26	129	37	94	28	346
Tropical, woody	32	9	102	27	133	35	109	29	376
Total	589		661		571		477		2298

woody groups have, in general, been derived from tropical ancestors.[225] Most conspicuous in this connection is the family Annonaceae, which is almost entirely tropical, and has chiefly the basic numbers $x = 7$, 8, and 9. In respect of floral anatomy, it is one of the most primitive families of angiosperms.[38]

The second line of evidence stems from the fact that many woody families or tribes have basic numbers much higher than the mode of $x = 12$, 13, or 14, which are almost certainly of polyploid origin. Those of the Rosaceae tribe Pomoideae ($x = 17$), Oleaceae tribe Oleoideae ($x = 23$), Salicaceae ($x = 19$), Magnoliaceae ($x = 19$), Platanaceae ($x = 21$), and the genera *Tilia* ($x = 41$), *Aesculus* ($x = 20$) and *Erythrina* ($x = 21$) have long been known.[42] Recently, many other such records have been added,[66] such as genera of Winteraceae ($x = 43$) and Bombacaceae ($x = 28, 36$). The existence of these groups, many of which are wide-spread and highly diversified in both temperate and tropical regions, shows that extensive polyploidy must have taken place during the early evolution of woody angiosperms. Hence a plausible assumption is that this series of polyploid cycles began with the lowest basic numbers now found in woody angiosperms, and that the present rarity of species having these low numbers is due to the differential extinction of primitive diploids.

Were these ancient cycles of polyploidy among woody plants promoted by radical disturbances of the habitat and the occupation of newly available habitats, as were the more recent cycles of herbaceous perennials? This hypothesis, which has been favoured by those cytologists who have been particularly concerned with tropical floras,[66,154] has much to recommend it. The kinds of disturbances which could have existed during the early evolutionary history of angiosperms have already been discussed. The most probable general hypothesis, therefore, is that the polyploidy which gave rise to the basic numbers of woody plants took place at various times during the Cretaceous and the earliest part of the Tertiary period, while the diversification of species on the basis of secondary basic numbers is largely a product of the Tertiary and Quaternary periods.

Since many herbaceous genera of tropical regions also have high basic numbers, and low percentages of polyploidy within a genus, a reasonable hypothesis is that these groups also became differentiated as to genera during the Cretaceous period. This hypothesis is supported by the presence in both the Old World and the New World of many tropical and subtropical herbaceous genera, such as *Desmodium* and *Phaseolus* (Leguminosae), *Solanum*, *Begonia*, *Cyperus*, and *Oryza*. Whatever hypothesis might be invoked to explain the pantropical distribution of these genera, their extensive development on both hemispheres is hard to explain unless we assume that they had acquired their present wide distribution at least by the middle of the Tertiary period. Consequently, whatever conditions promoted the earlier cycles of polyploidy must have affected entire floras, including both woody and herbaceous species.

POLYPLOIDY AND THE ORIGIN OF HIGHER CATEGORIES

Cytologists have often debated the question: Does polyploidy produce only new variations on old themes, or can it be responsible for major evolutionary advances? The principal argument in favour of the latter point of view has been the existence of entire tribes, subfamilies or families, such as those mentioned in the last section, which have basic numbers of obvious polyploid derivation. Evidence favouring the former point of view is that most of the more advanced families of flowering plants, such as Compositae and Gramineae, contain many genera having basic numbers which are not of polyploid origin. Where genera with high basic numbers exist, they are most easily explained as the polyploid derivatives of genera having low basic numbers which already possessed the morphological specializations of the family in question. More recently, the case against polyploidy as a progressive agent has been strengthened by the increasing number of chromosome counts in woody genera of various families which are so low that they cannot be of polyploid origin. There is now good reason to believe that when chromosome numbers of the majority of tropical woody genera become known, the phylogenetic sequences of families which include genera having original diploid basic numbers will be nearly as complete as those based upon the morphology of existing forms. The existence of such sequences would indicate that polyploidy has been important in the diversification of genera and species within families, but not in the origin of the families and orders themselves.

Another highly significant fact is that both among angiosperms and spore-bearing vascular plants there are much greater differences in basic number between different primitive groups than between more advanced ones. The range of chromosome numbers among genera and families of woody Ranales, which are now generally recognized as the most primitive angiosperms, is illustrated in the diagram, Figure 6.21. Figure 6.22 also suggests how the higher numbers could have arisen. The highest of these numbers ($x = 43$) would have to be regarded as 12-ploid in comparison to the lowest ($x = 7$). Both the low chromosome numbered (Annonaceae) and high numbered (Winteraceae) groups are regarded as very primitive on the basis of their morphology and anatomy; and the latter are usually considered to be the more primitive of the two. In contrast, the more advanced woody families, such as Apocynaceae, Rubiaceae, Ulmaceae, Moraceae, and Fagaceae, all have relatively homogeneous and similar chromosome numbers, falling into the modal range of $x = 11, 12, 13$, and 14.

The distribution of chromosome numbers in angiosperms as a whole,

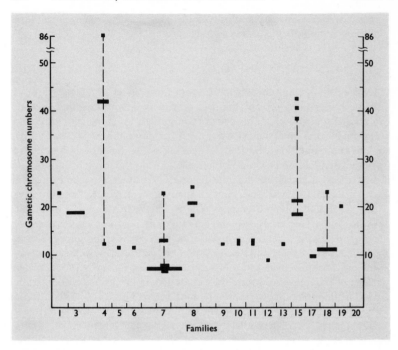

Fig. 6.21 Diagram showing the distribution of chromosome numbers in the families of the primitive order Magnoliales, as recognized by Cronquist.[38] Numbers on the horizontal axis are those of the families as he lists them: 1. Austrobaileyaceae; 3. Magnoliaceae; 4. Winteraceae; 5. Degeneriaceae; 6. Himantandraceae; 7. Annonaceae; 8. Myristicaceae; 9. Canellaceae; 10. Illiciaceae; 11. Schisandraceae; 12. Eupomatiaceae; 13. Amborellaceae; 15. Monimiaceae; 17. Calycanthaceae; 18. Lauraceae; 19. Hernandiaceae. Families nos. 2 (Lactoridaceae), 14 (Trimeniaceae) and 16 (Gomortegaceae) are omitted, since they are unknown cytologically. Each square or bar represents, horizontally, the number of genera having basic numbers as indicated.

therefore, suggests that there was a great burst of polyploidy, probably associated with hybridization, during the initial diversification and expansion of the group. Afterward, most of the diploid ancestors of these polyploids became extinct, but before they did so they gave rise to various herbaceous diploid groups, from which the more recent polyploid series have arisen. Many of the high polyploids survived as relictual forms, but did not participate in major evolutionary trends. On the other hand, woody polyploids at intermediate levels, particularly those arising from diploids which had already acquired a considerable degree of morphological specialization, spread, diversified, and gave rise to the rich flora of woody angiosperms which characterizes modern forests, particularly in the tropics.

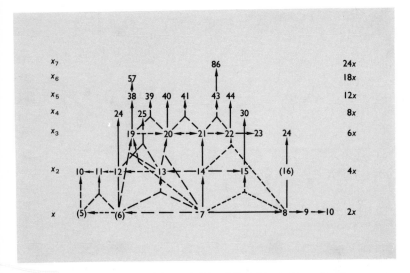

Fig. 6.22 An interpretation of the probable relationships between the chromosome numbers represented in Figure 6.21. Continuous lines (———) represent reasonably certain phylogenetic connections; coarse broken lines (— —) represent probable connections; fine broken lines (– – – –) represent possible connections. (From Ehrendorfer et al.[66])

The distribution of basic numbers in spore-bearing vascular plants is similar, except that much higher numbers are involved. In the class Lycopsida, which are generally regarded as primitive, both low ($n = 7$) and very high ($n = 132$) chromosome numbers exist[113,150] while in the most advanced group of ferns the numbers, though mostly high, vary less from one genus to another.

Apparently, therefore, the best generalization which can be made to fit the known facts concerning chromosome numbers and morphological specialization is that of *inverse correlation*. When families and orders are compared, the greatest variation in chromosome number is found in different primitive groups, while morphological specialization is in general associated with greater stabilization of the chromosome number. Great cytological diversity is associated with morphological conservatism, while morphological specialization is associated with cytological homogeneity or with alterations of individual chromosomes. This inverse correlation is best explained by the reasonable assumption that morphological specialization, as well as the evolution of new ways of becoming adapted to the environment, are based upon changes at the level of the gene. Multiplication of chromosome sets either has little effect upon evolutionary progress at the gene level, or actually tends to retard it.

References

1. ALSTON, R. E. and TURNER, B. L. (1963). *Biochemical Systematics*. Prentice-Hall, Englewood Cliffs, New Jersey.
2. ANDERSON, E. (1954). *Ann. Mo. bot. Gdn*, **41**, 305.
3. ANDERSON, E. and SAX, K. (1936). *Bot. Gaz.*, **97**, 433.
4. AVDULOV, N. P. (1931). *Bull. appl. Bot. Genet. Pl. Breed. (Suppl.)*, **44**, 1.
5. AXELROD, D. I. (1958). *Bot. Rev.*, **24**, 433.
6. BABCOCK, E. B. (1947). *The Genus Crepis*, I. and II. University of California Press, Berkely.
7. BABCOCK, E. B. and STEBBINS, G. L. (1938). The American Species of Crepis. *Publs Carnegie Instn*, **504**.
8. BAETCKE, K. P., SPARROW, A. H., NAUMAN, C. H. and SCHWEMMER, S. S. (1967). *Proc. natn. Acad. Sci. U.S.A.*, **58**, 533.
9. BAKER, H. G. (1965). Characteristics and modes of origin of weeds. In *The Genetics of Colonizing Species*, Baker, H. G. and Stebbins, G. L., eds. Academic Press, New York.
10. BAKER, H. G. (1967). *Taxon*, **16**, 293.
11. BALDWIN, J. T. (1936). *J. Genet.*, **33**, 455.
12. BAMFORD, R. and WINKLER, F. B. (1941). *J. Hered.*, **32**, 218.
13. BATTAGLIA, E. (1964). *Caryologia*, **17**, 245.
14. BEETLE, A. A. (1960). A Study of Sagebrush. The Section Tridentatae of *Artemisia*. *Univ. Wyoming agric. Exp. Stn Bull.*, **368**.
15. BELL, C. R. (1954). *Univ. Calif. Publs Bot.*, **27**, 133.
16. BIRNSTIEL, M. L., WALLACE, H., SIRLIN, J. L. and FISCHBERG, M. (1966). *Natn. Cancer Inst. Monogr.*, **23**, 431.
17. BOCHER, T. W. (1960). *Biol. Skr.*, **11**, 1.
18. BOCHER, T. W. (1962). *Bot. Tidsskr.*, **58**, 1.
19. BONNER, J. et al. (1968). *Science, N.Y.*, **159**, 47.
20. BOSEMARK, N. O. (1956). *Hereditas*, **42**, 189.
21. BOWDEN, W. M. (1945). *Am. J. Bot.*, **32**, 191.
22. BRINK, R. A. (1964). *Am. Nat.*, **98**, 193.

23. BRITTEN, R. J. and KOHNE, D. E. (1968). *Science, N.Y.*, **161**, 529.
24. BROWN, D. D. and GURDON, J. B. (1964). *Proc. natn. Acad. Sci. U.S.A.*, **51**, 139.
25. BROWN, S. W. (1966). *Science, N.Y.*, **151**, 417.
26. BROWN, S. W. and ZOHARY, D. (1955). *Genetics*, **40**, 850.
27. BROWN, W. V. and EMERY, W. H. P. (1957). *Bot. Gaz.*, **118**, 246.
28. CHANEY, R. W. (1940). *Bull. geol. Soc. Am.*, **51**, 469.
29. CHENNAVEERAIAH, M. S. (1960). *Acta Horti gotoburg.*, **23**, 85.
30. CHUANG, T-L. and HU, W. W. L. (1963). *Bot. Bull. Acad. Sinica*, IV, **1**, 10.
31. CLAUSEN, J. (1952). *Proc. 6th int. Grassland Congr.*, 216.
32. CLAUSEN, J. (1954). *Caryologia*, **7**, 469.
33. CLAUSEN, J., KECK, D. D. and HIESEY, W. M. (1940). Experimental Studies on the Nature of Species. I. The Effect of Varied Environments on Western North American Plants. *Publs Carnegie Instn*, **520**.
34. CLAUSEN, J., KECK, D. D. and HIESEY, W. M. (1948). Experimental Studies on the Nature of Species. III. Environmental responses of climatic races of *Achillea*. *Publs Carnegie Instn*, **581**.
35. CLELAND, R. E. (1957). *Proc. int. Genet. Symp., 1956 (Japan)*, 5.
36. CLELAND, R. E. (1962). *Adv. Genetics*, **11**, 147.
37. CLELAND, R. E. (1964). *Proc. Am. phil. Soc.*, **108**, 88.
38. CRONQUIST, A. (1968). *The Evolution and Classification of Flowering Plants*. Houghton Mifflin, Boston.
39. DARLINGTON, C. D. (1937). *Recent Advances in Cytology*, 2nd edition. Blakiston's, Philadelphia.
40. DARLINGTON, C. D. (1963). *Chromosome Botany and the Origins of Cultivated Plants*. Hafner Publishing Co., New York, and George Allen and Unwin, London.
41. DARLINGTON, C. D. (1965). *Cytology*. J. and A. Churchill, London.
42. DARLINGTON, C. D. and WYLIE, A. P. (1955). *Chromosome Atlas of Flowering Plants*. George Allen and Unwin, London.
43. DAVIDSON, E. H., CRIPPA, M., KRAMER, F. R. and MIRSKY, A. E. (1966). *Proc. natn. Acad. Sci. U.S.A.*, **56**, 856.
44. DAWSON, C. D. R. (1941). *J. Genet.*, **42**, 49.
45. DAY, A. (1965). *El Aliso*, **6**, 25.
46. DE, D. N. (1961). *The Nucleus*, **4**, 1.
47. DE, D. N. (1968). *Proc. XII int. Congr. Genet.*, **2**, 72.
48. DEAN, D. S. (1959). *Am. Midl. Nat.*, **6**, 204.
49. DELAUNAY, L. (1926). *Z. Zellforsch. mikrosk. Anat.*, **4**, 338.
50. DEMPSTER, L. T. and EHRENDORFER, F. (1965). *Brittonia*, **17**, 289.
51. DE WET, J. M. J. (1954). *Am. J. Bot.*, **41**, 204.
52. DE WET, J. M. J. (1965). *Am. Nat.*, **99**, 167.
53. DE WET, J. M. J. (1968). *Evolution, Lancaster, Pa.*, **22**, 394.
54. DE WET, J. M. J. and HARLAN, J. R. (1966). *Am. J. Bot.*, **53**, 94.
55. DOBZHANSKY, T. (1951). *Genetics and the Origin of Species*, 3rd edition. Columbia University Press, New York.
56. DOGADKINA, N. A. (1941). *C. r. Acad. Sci. U.R.S.S.*, **30**, 166.
57. DOWRICK, V. P. J. (1956). *Heredity*, **10**, 219.
58. DUNFORD, M. P. (1964). *Am. J. Bot.*, **51**, 49.
59. DU PRAW, E. J. (1965). *Nature, Lond.*, **206**, 338.
60. EHRENDORFER, F. (1953). *Ost. bot. Z.*, **100**, 583.
61. EHRENDORFER, F. (1960). *Z. VererbLehre*, **91**, 400.
62. EHRENDORFER, F. (1961). *Chromosomal*, **11**, 523.
63. EHRENDORFER, F. (1961). *Madrono*, **16**, 109.

64. EHRENDORFER, F. (1964). *Proc. XI int. Congr. Genet.*, The Hague, The Netherlands, 1963, 399.
65. EHRENFORFER, F. (1965). Dispersal mechanisms, genetic systems, and colonizing abilities in some flowering plant families. In *The Genetics of Colonizing Species*, Baker, H. G. and Stebbins, G. L., eds, 331. Academic Press, New York.
66. EHRENDORFER, F., KRENDL, F., HABELER, E. and SAUER, W. (1968). *Taxon*, **17**, 337.
67. EIG, A. (1936). Aegilops. In *Die Pflanzenareale*, Hannig, E. and Winkler, H., eds, **4**, 43.
68. ELLERSTROM, S. and HAGBERG, A. (1954). *Hereditas*, **40**, 535.
69. FAVARGER, C. (1957).*C. r. VIII Congr. Int. Bot. (Paris)*, **9-10**, 51.
70. FAVARGER, C. (1964). *Ber. dt. bot. Ges.*, **77**, 73.
71. FAVARGER, C. and CONTANDRIOPOULOS, J. (1961). *Bull. Soc. Bot. Suisse*, **71**, 384.
72. FEINBRUN, N. (1958). *Evolution, Lancaster, Pa.*, **12**, 173.
73. FELDHERR, C. M. (1965). *J. Cell Biol.*, **25**, 43.
74. FROST, S. (1962). *Hereditas*, **48**, 667.
75. FUKUDA, I. (1967). *Taxon*, **16**, 308.
76. FUKUDA, I. (1969). Unpublished data.
77. FURNKRANZ, D. (1965). *Ber. dt. bot. Ges.*, **78**, 139.
78. FURNKRANZ, D. (1966). *Ost. bot. Z.*, **113**, 427.
79. GADELLA, T. W. J. (1964). *Wentia*, **11**, 1.
80. GALL, J. (1966). *Natn. Cancer Inst. Monogr.*, **23**, 475.
81. GARBER, E. D. (1956). *Bot. Gaz.*, **118**, 71.
82. GERSTEL, D. U. and PHILLIPS, L. L. (1958). *Cold Spring Harb. Symp. quant. Biol.*, **23**, 225.
83. GOULIAN, M. and KORNBERG, A. (1967). *Proc. natn. Acad. Sci. U.S.A.*, **58**, 1723.
84. GRANT, A. and GRANT, V. (1956). *El Aliso*, **3**, 203.
85. GRANT, V. (1954). *El Aliso*, **3**, 19.
86. GRANT, V. (1966). *Genetics*, **53**, 757.
87. GRANT, W. F. (1955). *Brittonia*, **8**, 121.
88. GRANT, W. F. (1965). *Can. J. Genet. Cytol.*, **7**, 457.
89. GRANT, W. F. and ZANDSTRA, I. I. (1968). *Can. J. Bot.*, **46**, 585.
90. GRAZI, F., UMAERUS, M. and ÅKERBERG, E. (1961). *Hereditas*, **47**, 489.
91. GREENLEAF, W. H. (1938). *J. Hered.*, **29**, 451.
92. GREGORY, W. D. (1941). *Trans. Am. phil. Soc.*, n.s., **31**, 443.
93. GROSS, J. (1961). *Scient. Am.*, **204**, 121.
94. GUINOCHET, M. (1942). *Bull. Soc. bot. Fr.*, **89**, 153.
95. HAGA, T. and KURABAYASHI, M. (1953). *Cytologia*, **18**, 13.
96. HAIR, J. B. and BEUZENBERG, E. L. (1958). *Nature, Lond.*, **181**, 1584.
97. HARNEY, P. and GRANT, W. F. (1964). *Am. J. Bot.*, **51**, 621.
98. HECKARD, L. R. (1960). *Univ. Calif. Publs Bot.*, **32**, 1.
99. HECKARD, L. R. (1968). *Brittonia*, **20**, 212.
100. HEDBERG, I. (1967). *Symb. bot. upsal.*, **18**, 1.
101. HEITZ, E. (1929). *Ber. dt. bot. Ges.*, **47**, 274.
102. HEITZ, E. (1932). *Planta*, **18**, 571.
103. HESS, O. (1966). Structural Modifications of the Y-Chromosome in *Drosophila hydei* and their Relation to Gene Activity. In *Chromosomes Today*, Vol. 1, Darlington, C. D. and Lewis, K. R., eds, 167. Oliver and Boyd, Edinburgh.

104. HITCHCOCK, C. and KRUCKEBERG, A. R. (1957). *Univ. Wash. Publs Biol.* **18**, 1.
105. HOFFMAN, L. R. (1967). *Am. J. Bot.*, **54**, 271.
106. HUBAC, J. M. (1961). *Bull Soc. bot. Fr.*, **108**, 1.
107. HUTCHINSON, J. (1934). *The Families of Flowering Plants. II. Mono-cotyledons.* MacMillan, London.
108. HUZIWARA, Y. (1959). *Evolution, Lancaster, Pa.*, **13**, 188.
109. HYDE, B. B. (1953). *Am. J. Bot.*, **40**, 809.
110. INGRAM, V. I. (1963). *The Hemoglobins in Genetics and Evolution.* Columbia University Press, New York.
111. JACKSON, R. C. (1962). *Am. J. Bot.*, **49**, 119.
112. JACKSON, R. C. (1965). *Am. J. Bot.*, **52**, 946.
113. JERMY, A. C., JONES, K. and COLDEN, C. (1967). *J. Linn. Soc. (Bot.)*, **60**, 147.
114. JOHNSON, A. W. and PACKER, J. G. (1965). *Science, N.Y.*, **148**, 237.
115. JOHNSON, B. L. and HALL, O. (1965). *Am. J. Bot.*, **52**, 506.
116. JONES, K. (1964). *Chromosoma*, **15**, 248.
117. JONES, K. (1966). *Kew Bull.*, **20**, 357.
118. JONES, K. (1967). *Kew Bull.*, **20**, 151.
119. JONG, K. and LETHBRIDGE, A. (1967). *Notes R. bot. Gdn Edinb.*, **27**, 175.
120. KACHIDZE, N. (1929). *Planta*, **7**, 482.
121. KANEKO, K. and MAEKAWA, F. (1968). *J. Fac. Sci. Tokyo Univ., Sect.* III, **10**, 1.
122. KHOSHOO, T. N. and AHUJA, M. R. (1963). *Chromosoma*, **14**, 522.
123. KIHARA, H. (1954). *Cytologia*, **19**, 336.
124. KNOX, R. B. (1967). *Science, N.Y.*, **157**, 6.
125. KNOX, R. B. and HESLOP-HARRISON, J. (1963). *Bot. Notiser*, **116**, 127.
126. KURABAYASHI, M. (1958). *Evolution, Lancaster, Pa.*, **12**, 286.
127. KURABAYASHI, M., LEWIS, H. and RAVEN, P. H. (1962). *Am. J. Bot.*, **49**, 1003.
128. KURITA, M. (1960). *Mem. Ehime Univ., Sect.* II *(Sci.), Ser.* B *(Biol.)* **4**, 53.
129. KYHOS, D. W. (1965). *Evolution, Lancaster, Pa.*, **19**, 26.
130. LANCE, A. (1957). *Annls Sci. nat. (Bot.), 11e sér.*, **18**, 91.
131. LEE, C. L. (1954). *Am. J. Bot.*, **41**, 545.
132. LEVAN, A. (1935). *Hereditas*, **20**, 289.
133. LEVAN, A. (1939). *Hereditas*, **25**, 109.
134. LEVITZKY, G. A. (1931). *Bull. appl. Bot. Genet. Pl. Breed.*, **27**, 220.
135. LEWITZKY (LEVITZKY), G. and SIZOWA, M. (1934). *C. r. Acad. Sci. U.R.S.S.*, **1934**, 86.
136. LEWIS, H. (1953). *Evolution, Lancaster, Pa.*, **7**, 1.
137. LEWIS, H. (1966). *Science, N.Y.*, **152**, 167.
138. LEWIS, H. and EPLING, C. (1959). *Evolution, Lancaster, Pa.*, **13**, 511.
139. LEWIS, H. and LEWIS, M. E. (1955). *Univ. Calif. Publs Bot.*, **20**, 241.
140. LEWIS, H. and RAVEN, P. H. (1958). *Evolution, Lancaster, Pa.*, **12**, 319.
141. LEWIS, H. and SZWEYKOWSKI, J. (1964). *Brittonia*, **16**, 343.
142. LEWIS, W. H., OLIVER, R. L. and SUDA, Y. *Ann. Mo. bot. Gdn*, **54**, 153.
143. LIMA-DE-FARIA, A. (1954). *Chromosoma*, **6**, 330.
144. LIMA-DE-FARIA, A. (1956). *Hereditas*, **42**, 85.
145. LIMA-DE-FARIA, A. (1956). *Cytologia, Suppl. Vol. Proc. int. Genet. Symp.* 108.
146. LIMA-DE-FARIA, A. (1959). *J. biophys. biochem. Cytol.*, **6**, 457.

147. LIMA-DE-FARIA, A., SARVELLA, P. and MORRIS, R. (1959). *Hereditas*, **45**, 467.
148. LOEWENSTEIN, W. R., KANNO, Y. and ITO, S. (1966). *Ann. N. Y. Acad. Sci.*, **137**, 708.
149. LONGLEY, A. E. (1938). *J. agric. Res.*, **56**, 177.
150. LOVE, A. and LOVE, D. (1958). *The Nucleus*, **1**, 1.
151. LOVE, A. and SARKAR, N. (1956). *Can. J. Bot.*, **34**, 261.
152. LOVE, R. M. (1954). *Am. J. Bot.*, **41**, 107.
153. LOVELL, J. H. (1918). *The Flower and the Bee.* Scribner's, New York.
154. MANGENOT, S. and MANGENOT, G. (1962). *Revue Cytol. Biol. vég.*, **25**, 411.
155. MANTON, I. (1934). *Z. indukt. Abstamm.- u. VererbLehre*, **67**, 41.
156. MANTON, I. (1950). *Problems of Cytology and Evolution in the Pteridophyta.* Cambridge University Press.
157. MANTON, I. and SLEDGE, W. A. (1954). *Phil. Trans. R. Soc.*, **238**, 125.
158. MARCHANT, C. J. (1967). *Kew Bull.*, **21**, 161.
159. MARCHANT, C. J. (1968). *Chromosoma*, **24**, 100.
160. MARKS, G. E. (1956). *New Phytol.*, **55**, 120.
161. MARKS, G. E. (1957). *Chromosoma*, **8**, 650.
162. MARKS, G. E. (1966). *Evolution, Lancaster, Pa.*, **20**, 552.
163. MAYR, E. (1963). *Animal Species and Evolution.* Harvard University Press, Cambridge, Mass., U.S.A.
164. MCCARTHY, B. J. and HOYER, B. H. (1964). *Proc. natn. Acad. Sci. U.S.A.*, **52**, 915.
165. MOORE, D. M. and LEWIS, H. (1966). *Heredity*, **21**, 37.
166. MORTON, J. K. (1962). *J. Linn. Soc. (Bot.)*, **48**, 231.
167. MORTON, J. K. (1967). *J. Linn. Soc. (Bot.)*, **60**, 167.
168. MOSQUIN, Th. (1964). *Evolution, Lancaster, Pa.*, **18**, 12.
169. MOSQUIN, Th. (1967). *Evolution, Lancaster, Pa.*, **21**, 713.
170. MULLIGAN, G. A. (1960). *Can. J. Genet. Cytol.*, **2**, 150.
171. MUNTZING, A. (1958). *Hereditas*, **44**, 145.
172. MUNTZING, A. (1958). *Proc. X int. Cong. Genet.*, **1**, 453.
173. NAGL, W. (1962). *Naturwissenschaften*, **49**, 2.
174. NINAN, C. A. (1958). *Cytologia*, **23**, 291.
175. NODA, S. (1966). *Bull. Osaka Gakuin Univ.*, **6**, 85.
176. NORDENSKIOLD, H. (1951). *Hereditas*, **37**, 324.
177. OWNBEY, M. (1950). *Am. J. Bot.*, **37**, 487.
178. PAYNE, W. W., RAVEN, P. H. and KYHOS, D. W. (1964). *Am. J. Bot.*, **51**, 419.
179. PERRY, R. P. (1962). *Proc. natn. Acad. Sci. U.S.A.*, **48**, 2179.
180. RAO, R. S., KAMMATHY, R. V. and RAGHAVAN, R. S. (1968). *J. Linn. Soc. (Bot.)*, **60**, 357.
181. RAVEN, P. H. and LEWIS, H. (1959). *Brittonia*, **11**, 193.
182. RAVEN, P. H., SOLBRIG, O. T., KYHOS, D. W. and SNOW, R. (1960). *Am. J. Bot.*, **47**, 124.
183. RAY P. M. and CHISAKI, F. (1957). *Am. J. Bot.*, **44**, 545.
184. RICHARDSON, M. M. (1936). *J. Genet.*, **32**, 411.
185. RILEY R. (1966). *Scientific Progress*, **54**, 193.
186. RITOSSA, F. M. and ATWOOD, K. C. (1966). *Proc. natn. Acad. Sci. U.S.A.*, **56**, 496.
187. RITOSSA, F. M., ATWOOD, K. C. and SPIEGELMAN, S. (1966). *Genetics*, **54**, 819.
188. ROTHWELL, N. V. and KUMP, J. G. (1965). *Am. J. Bot.*, **52**, 403.
189. SAMEJIMA, J. (1958). *Cytologia*, **23**, 159.

190. SCANDALIOS, . (1969). *Biochem. Genet.*, **3**, 37.
191. SCHNEIDER, I. (1958). *Ost. bot. Z.*, **105**, 112.
192. SCHULZE, K. L. (1939). *Arch. Protistenk.*, **92**, 179.
193. SEGAWA, M. (1965). *J. Sci. Hiroshima Univ., Ser. B, Div. 2 (Bot.)*, **10**, 149.
194. SIMPSON, G. G. (1953). *The Major Features of Evolution.* Columbia University Press, New York.
195. SINSHEIMER, R. L. (1957). *Science, N.Y.*, **125**, 1123.
196. SMALL, E. (1968). *Brittonia*, **20**, 169.
197. SMITH, B. W. (1967). *Evolution, Lancaster, Pa.*, **18**, 93.
198. SMITH, B. W. (1968). *Am. J. Bot.*, **55**, 673.
199. SNOAD, B. (1951). *Heredity*, **5**, 279.
200. SODERSTROM, T. R. and BEAMAN, J. H. (1968). *Publs Mich. St. Univ. Mus. Biol. Ser.*, **3**, 465.
201. SOLBRIG, O. T. (1960). *Contr. Gray Herb. Harv.*, **188**, 1.
202. SPARROW, A. J. and EVANS, H. J. (1961). *Brookhaven Symp. Biol.*, **14**, 76.
203. STAUDT, G. (1953). *Ber. dt. bot. Ges.*, **66**, 237.
204. STAUDT, G. (1967). *Z. PflZücht.*, **58**, 245.
205. STAUDT, G. (1967). *Z. PflZücht.*, **58**, 309.
206. STEBBINS, G. L. (1938). *Genetics*, **23**, 83.
207. STEBBINS, G. L. (1938). *Am. J. Bot.*, **25**, 189.
208. STEBBINS, G. L. (1942). *Am. Nat.*, **76**, 36.
209. STEBBINS, G. L. (1950). *Variation and Evolution in Plants.* Columbia University Press, New York.
210. STEBBINS, G. L. (1956). Artificial Polyploidy as a Tool in Plant Breeding. In *Brookhaven Symp. Biol.-Genet. in plant breeding*, Smith, H., ed., **9**, 37.
211. STEBBINS, G. L. (1956). *Am. J. Bot.*, **43**, 890.
212. STEBBINS, G. L. (1958). *Cold Spring Harb. Symp. quant. Biol.*, **23**, 365.
213. STEBBINS, G. L. (1959). *Proc. Am. phil. Soc.*, **103**, 231.
214. STEBBINS, G. L. (1959). Genes, Chromosomes, and Evolution. In *Vistas in Botany, I*, Turrill, W. B., ed., 258. Pergamon Press, London.
215. STEBBINS, G. L. (1965). Colonizing Species of the Native California Flora. In *The Genetics of Colonizing Species*, Baker, H. G. and Stebbins, G. L., eds, 173. Academic Press, New York.
216. STEBBINS, G. L. (1966). *Processes of Organic Evolution.* Prentice-Hall, Englewood Cliffs, New Jersey.
217. STEBBINS, G. L. (1966). *Science, N.Y.*, **152**, 1463.
218. STEBBINS, G. L. and DAY, A. (1967). *Evolution, Lancaster, Pa.*, **21**, 409.
219. STEBBINS, G. L., JENKINS, J. A. and WALTERS, M. S. (1953). *Univ. Calif. Publs Bot.*, **26**, 401.
220. STEBBINS, G. L. and MAJOR, J. (1965). *Ecol. Monogr.*, **35**, 1.
221. STEBBINS, G. L., TOBGY, H. A. and HARLAN, J. R. (1944). *Proc. Calif. Acad. Sci.*, **25**, 307.
222. STEBBINS, G. L., VALENCIA, J. I. and VALENCIA, R. A. (1946). *Am. J. Bot.*, **33**, 338.
223. STEBBINS, G. L. and ZOHARY, D. (1959). *Univ. Calif. Publs Bot.*, **31**, 1.
224. STEINER, E. (1956). *Genetics*, **41**, 486.
225. TAKHTAJAN, A. (1969). *Flowering Plants: Origin and Dispersal.* Oliver and Boyd, Edinburgh.
226. TANAKA, R. (1965). *Bot. Mag.*, **78**, 50.
227. TATUNO, S. (1957). *J. Sci. Hiroshima Univ., Ser. B, Div. 2 (Bot.)*, **8**, 81.
228. TAYLOR, R. L., MARCHAND, L. S. and CRAMPTON, C. W. (1964). *Can. J. Genet. Cytol.*, **6**, 42.

229. TAYLOR, R. L. and MULLIGAN, G. A. (1968). *Research Branch Can. Dept. Agric. Monogr.*, **44**, Pt. 2, 1.
230. THOMPSON, H. J. (1953). *Contr. Dudley Herb.*, **4**, 73.
231. TOBGY, H. A. (1943). *J. Genet.*, **45**, 67.
232. TSCHERMAK-WOESS, E. (1957). *Chromosoma*, **8**, 523.
233. TSCHERMAK-WOESS, E. and HASITSCHKA, H. (1954). *Öst. bot. Z.*, **101**, 79.
234. UHL, H. and MORAN, R. (1953). *Am. J. Bot.*, **40**, 492.
235. UNDERBRINK, A. G., TING, Y. C. and SPARROW, A. L. (1967). *Can. J. Genet. Cytol.*, **9**, 606.
236. UPCOTT, M. (1939). *J. Genet.*, **39**, 79.
237. WARD, G. H. (1953). *Contr. Dudley Herb.*, **4**, 155.
238. WATKINS, G. M. (1936). *Am. J. Bot.*, **23**, 328.
239. WEBBER, J. M. (1932). *Am. J. Bot.*, **19**, 411.
240. WESTERGAARD, M. (1958). *Adv. Genet.*, **9**, 217.
241. WESTERGAARD, M. and von WETTSTEIN, D. (1966). *C. r. Trav. Lab. Carlsberg*, **35**, 261.
242. WHITE, M. J. D. (1954). *Animal Cytology and Evolution.* Cambridge University Press.
243. WIENS, D. (1964). *Am. J. Bot.*, **51**, 1.
244. WILKINS, M. H. F. (1963). *Science, N.Y.*, **140**, 941.
245. WILLIS, J. C. (1922). *Age and Area.* Cambridge University Press.
246. WOLFE, S. L. and MARTIN, P. G. (1968). *Exp. Cell Res.*, **50**, 140.
247. WOOD, C. E. JR. (1955). *Rhodora*, **57**, 105.
248. WOODARD, J., GOROVSKY, M. and SWIFT, H. (1966). *Science, N.Y.*, **151**, 215.
249. WULFF, H. D. (1939). *Ber. dt. bot. Ges.*, **57**, 84.
250. ZANDSTRA, I. I. and GRANT, W. F. (1967). *Can. J. Bot.*, **46**, 557.
251. ZOHARY, D. (1965). Colonizer Species in the Wheat Group. In *The Genetics of Colonizing Species*, Baker, H. G. and Stebbins, G. L., eds, 403. Academic Press, New York.
252. ZOHARY, D. and NUR, U. (1959). *Evolution, Lancaster, Pa.*, **13**, 311.
253. ZUK, J. (1969). Autoradiographic Studies in *Rumex* with special reference to sex chromosomes. In *Chromosomes Today*, Vol. 2, Darlington, C. D. and Lewis, K. R., eds, 183. Oliver and Boyd, Edinburgh.

Index